Exploring Science

Grade **8**

Derek McMonagle

Reviewers: Marlene Grey-Tomlinson, Bernadette Ranglin, Maxine McFarlane & Monacia Williams

Collins

William Collins' dream of knowledge for all began with the publication of his first book in 1819.

A self-educated mill worker, he not only enriched millions of lives, but also founded a flourishing publishing house. Today, staying true to this spirit, Collins books are packed with inspiration, innovation and practical expertise. They place you at the centre of a world of possibility and give you exactly what you need to explore it.

Collins. Freedom to teach.

Published by Collins
An imprint of HarperCollins*Publishers*
The News Building
1 London Bridge Street
London
SE1 9GF

HarperCollins*Publishers* Macken House
39/40 Mayor Street Upper
Dublin 1
DO1 C9W8
Ireland

Browse the complete Collins Caribbean catalogue at
www.collins.co.uk/caribbeanschools

10 9 8 7 6 5

ISBN 978-0-00-826328-7

British Library Cataloguing in Publication Data
A catalogue record for this publication is available from the British Library.

Publisher: Elaine Higgleton
Commissioning editor: Tom Hardy
In-house senior editor: Julianna Dunn
Author: Derek McMonagle
Reviewers: Marlene Grey-Tomlinson, Bernadette Ranglin, Maxine McFarlane & Monacia Williams
Project manager: Alissa McWhinnie, QBS Learning
Copyeditor: Rebecca Ramsden & Mitch Fitton
Proofreader: Sara Hulse & David Hemsley
Photo researcher, illustrator and typesetter: QBS Learning
Cover designer: Gordon MacGilp
Series designer: Kevin Robbins
Cover photo: think4photop/Shutterstock
Production controller: Tina Paul
Printed and bound by Ashford Colour Press Ltd

The publishers gratefully acknowledge the permission granted to reproduce the copyright material in this book. Every effort has been made to trace copyright holders and to obtain their permission for the use of copyright material. The publishers will gladly receive any information enabling them to rectify any error or omission at the first opportunity.

MIX
Paper | Supporting responsible forestry
FSC™ C007454

This book contains FSC™ certified paper and other controlled sources to ensure responsible forest management.

For more information visit: www.harpercollins.co.uk/green

Contents

Introduction – How to use this book   4

Unit 1:   Working like a scientist   6

Unit 2:   Photosynthesis and energy relationships   18

Unit 3:   More about matter   40

Unit 4:   Human nutrition   64

Unit 5:   Physical and chemical changes   90

Unit 6:   Forces and motion   122

Unit 7:   Respiration and gas exchange   152

Unit 8:   Space science   174

Unit 9:   Water and the Earth's atmosphere   204

Index   226

Acknowledgements   233

Introduction – How to use this book

Grade 8

Structure of a leaf

We are learning how to:

- describe the structure of a leaf
- appreciate how a leaf is designed to carry out photosynthesis.

Structure of a leaf

Any part of a plant that is green is able to carry out photosynthesis but the majority of glucose production in a plant takes place in the leaves.

Leaves come in different shapes and sizes but, with a few exceptions, they always have a large surface area compared to their volume. This is an advantage because it allows each leaf to absorb as much sunlight as possible. The total area of leaf on a large tree is similar to the surface area of a football pitch.

In addition to sunlight and chlorophyll, in order for photosynthesis to take place a leaf must be supplied with carbon dioxide gas and water. Carbon dioxide gas is able to enter the leaf through openings called stomata (singular **stoma**). These are found mostly on the underside of the leaf. The oxygen produced during photosynthesis passes out of the leaf in the same way. A plant obtains water from the ground. It is absorbed through the plant roots and passes up through the stem to branches, and eventually to individual leaves.

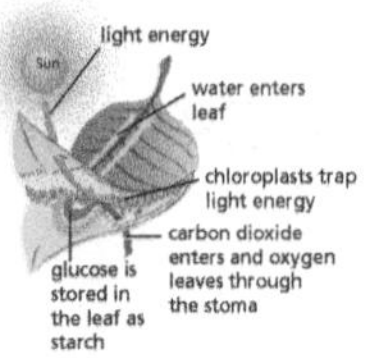

The initial product of photosynthesis is a **carbohydrate** called **glucose**. This chemical is very soluble in water. As water passes into and out of cells, glucose would dissolve in it, therefore it is not possible for a plant to store carbohydrate as glucose.

Glucose consists of small molecules. Many of these combine to form large molecules of another carbohydrate called **starch**. Starch is an example of a **polymer**. This is a large molecule formed by joining together many small molecules. Starch is much less soluble in water and can be stored in plant cells. Starch is sometimes described as a storage polymer.

FIG 2.2.1 Leaves come in different shapes and sizes

light energy
sun
water enters leaf
chloroplasts trap light energy
carbon dioxide enters and oxygen leaves through the stoma
glucose is stored in the leaf as starch

FIG 2.2.2 Inside a leaf

Activity 2.2.1

Testing a leaf for starch

Here is what you need:

- Green leaf
- Tweezers
- Test tube
- White tile
- Beaker of hot water
- Ethanol
- Iodine solution
- Pipette.

20   Exploring Science Grade 8: Unit 2: Photosynthesis and energy relationships

Here is what you should do:

1. Using tweezers hold the leaf in water bath (beaker with hot water) for about 1 minute. Observe what happens to the leaf.
2. Fold the leaf and place it in a test tube. Pour enough ethanol into the test tube to cover the leaf.
3. Stand the test tube in hot water bath for about 10 minutes. Give the test tube an occasional shake. Observe what happens to the leaf.
4. Remove the test tube from the water bath.
5. Take the leaf out of the test tube. Gently touch the leaf to see how it feels.
6. Wash or dip the leaf in cold water. Gently touch the leaf to see how it feels.
7. Spread the leaf on a tile and add a few drops of iodine solution using a dropper.
8. What colour does the leaf become? What conclusions can you draw from this?

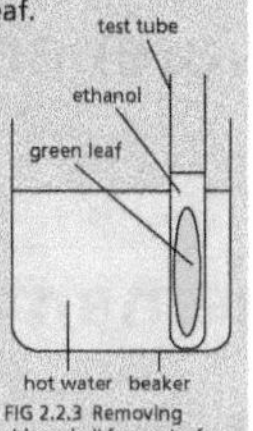

test tube
ethanol
green leaf
hot water   beaker
FIG 2.2.3 Removing chlorophyll from a leaf

A simple way to test if photosynthesis has taken place is by testing a leaf for the presence of starch. When iodine solution is added to starch, the mixture turns a characteristic blue-black colour. In order to see the colour change clearly, chlorophyll is usually removed from the leaf by warming in ethanol.

2.2

Fun fact

Iodine is not very soluble in water. In order to make a suitable solution to test for starch the iodine is dissolved in potassium iodide solution.

Key terms

stoma (plural stomata) small opening, found mostly on the underside of a leaf

carbohydrate substance containing carbon and hydrogen and oxygen in the same ratio as water, 2:1

glucose carbohydrate formed during photosynthesis

starch form of carbohydrate stored in plants

polymer a large molecule formed by adding many small molecules of the same type

Check your understanding

1. A student investigated the effects of temperature and light intensity on the rate of photosynthesis in tomato plants. The results were used to plot the following graphs.

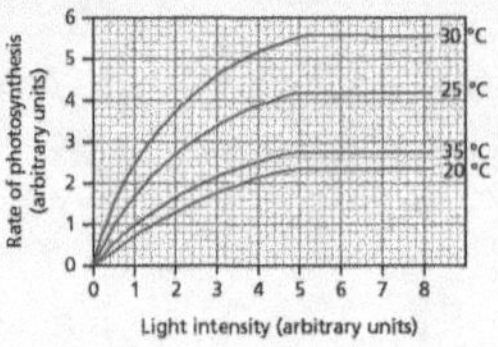

FIG 2.2.4

a) Why do all of the curves flatten out after about 5 units of light?

b) Under maximum light intensity, at which temperature was the rate of photosynthesis:

i) highest?   ii) lowest?

21

This tells you what you will be learning about in this lesson.

This introduces the topic.

The book has plenty of good illustrations to put the science into context.

There are often some fascinating fun facts

Each spread has activities to help you to investigate the topic.

Each spread offers questions to help you to check whether you have understood the topic.

Key terms are defined on the pages where they are used.

Review of Photosynthesis and energy relationships

- Photosynthesis is the process by which green plants trap energy from sunlight and use it to make food.
- During photosynthesis carbon dioxide and water are used up and glucose and oxygen are formed.
- Glucose and starch are examples of carbohydrates.
- Food is stored in plants as starch.
- Starch turns iodine solution blue-black.
- All life on Earth is dependent on photosynthesis. Plants use the food directly. Animals either eat the plants (herbivores) or eat the animals that eat plants (carnivores).
- Green plants are called producers because they use energy from sunlight to produce the food.
- Animals are called consumers because they consume the food made by green plants.
  - Primary consumers eat the plants directly.
  - Secondary consumers eat the animals that eat plants.
- Decomposers break down organic material, releasing nutrients back into the ecosystem for use again by living organisms.
- The relationship between producers and consumers can be shown as:

  Producers ⟶ Consumers

  When the names of organisms are included the result is called a food chain.
- The arrow joining groups of organisms in a food chain can be interpreted as:
  - 'is eaten by'
  - transfer of energy
  - transfer of nutrients.
- All food chains:
  - start with a producer
  - are driven by energy from the Sun.
- A food web shows the feeding relationship between a number of different organisms and is a combination of a number of food chains.
- Changes which affect the population of one organism in a food chain or a food web will affect the populations of other organisms in some way.
- Human activities often affect organisms in food chains.

At the end of each group of units there are pages which list the key topics covered in the units. These will be useful for revision.

At the end of each section there are special questions to help you and your teacher review your knowledge, see if you can apply this knowledge to new situations and if you can use the science skills that you have developed.

Science, Technology, Education, Arts and Mathematics (STEAM) activities are included, which present real-life problems to be investigated and resolved using your science and technology skills. These pages are called **Science in practice**.

Review questions on Photosynthesis and energy relationships

1. a) Copy and complete the word equation for photosynthesis.

.... + water $\xrightarrow{\text{chlorophyll}}$ glucose + ...

b) Describe a test for the presence of starch in a leaf.

c) The plant shown in Fig 2.RQ.1 is unusual in that part of the stem and the leaves attached to it are white, while the rest of the plant is green.

Predict whether a leaf from the white part of the stem and a leaf from the green part of the stem would both give a positive test for starch. Explain your answer.

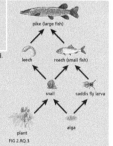

☐ leaf/stem white
☐ leaf/stem green
plant pot
soil

FIG 2.RQ.1

2. Fig 2.RQ.2 shows a food chain.

weeds   tadpoles   minnows   perch

FIG 2.RQ.2

a) Which organism is the primary consumer?

b) Which two animals in the food chain are carnivores?

c) State three ways in which the arrows in the food chain can be interpreted.

3. Fig 2.RQ.3 shows part of a food web for a pond.

a) What is the source of the energy that drives this food web?

b) Name a secondary consumer in the food web.

c) From this food web, write a food chain that contains at least three organisms.

pike (large fish)
leech   roach (small fish)
snail   caddis fly larva
plant   alga
FIG 2.RQ.3

Photosynthesis and energy relationships

Animals seldom have only one source of food. They may eat different foods according to season and/or availability. It is for this reason that an animal may feature on several different food chains in an ecosystem. Food chains can be combined to form a food web. A food web gives a more complete picture of the feeding relationships between organisms in an ecosystem.

Identify a small piece of land either within your school compound or close to your school or home. This land should be undeveloped and contain different types of plants.

FIG 2.SIP.1 A small piece of land

You may assume that there are plans to build a path on this land and install a bench. This will allow students or local people to sit and enjoy the surroundings.

The Nature Conservancy of Jamaica is concerned that this work will drastically affect the organisms that live on and near this land. They have asked you to construct a food web involving these organisms in order to better assess the impact of any changes. They are hoping that your report will allow them to identify the position for the path and bench that will cause least disturbance.

1. You are going to work in groups of 3 or 4 to investigate the feeding relationships between organisms on this land. The tasks are:
- To review the significance of food chains and food webs and how to construct them.
- To carry out research into some of the plants and animals you can expect to find in and around the land so that you will be able to identify them. You will also need to know something about the feeding habits of the animals so you can identify likely food sources.
- To spend time observing the organisms on and around the land and gathering information about their feeding habits.
- To construct a provisional food web for the land.
- To carry out further observations on the land to determine how accurate your food web is, and to supplement it with further observations.
- To compile a report, including a PowerPoint presentation, describing how you gathered the information needed to construct a food web. You should be prepared

Unit 1: Working like a scientist

We are learning how to:

- interpret data
- represent data as a pie chart.

Data and pie charts »

Data

Data is another word for information. Scientists often collect data of different kinds and display it in appropriate ways.

- Data may result from qualitative observations like colour or state.

- Data may result from quantitative observations like measurements of mass, volume or temperature.

FIG 1.1.1 Scientists at work

Readings should be recorded in a frequency table using suitable headings. Use a tally to count the **frequency** of each value.

In Table 1.1.1 the first 20 elements are classified as metals, non-metals or semi-metals (you will learn more about them in Unit 3). This is called a frequency table.

Classification	Elements	Tally	Number
Metals	lithium, beryllium, aluminium, sodium, magnesium, potassium, calcium	ℍℍ II	7
Non-metals	hydrogen, helium, carbon, nitrogen, oxygen, fluorine, neon, phosphorus, sulfur, chlorine, argon	ℍℍ ℍℍ I	11
Semi-metals	boron, silicon	II	2
		Total	20

TABLE 1.1.1

The tally is made by counting the number of elements in each group. The total should be 20 because we are concerned with the first 20 elements.

In Grade 7 you learned how to display the data in a frequency table as a bar graph or a line graph. Where the number of groups of data is small it is sometimes better to display the data in other ways.

Pie charts

For mathematical purposes pies are always round. A **pie chart** divides a pie into slices or sectors. The problem we have is deciding how big a slice or sector is needed to represent each group. This will depend on its frequency.

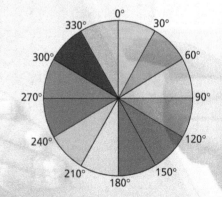

FIG 1.1.2 There are 360° at the centre of a circle

In Table 1.1.1 there are 20 elements. We want to represent 20 elements by 360°, therefore each element can be shown by a sector with an angle of $\frac{360}{20} = 18°$.

The number of degrees needed to represent each group of elements will depend on the frequency.

Group	Frequency	Angle at centre of sector
Metals	7	7 × 18 = 126°
Non-metals	11	11 × 18 = 198°
Semi-metals	2	2 × 18 = 36°
Totals	20	360°

TABLE 1.1.2

You should always check that the total of the angles at the centres of the sectors adds up to 360°. If it does not, you have made an error in your calculations. The pie chart for the first 20 elements will be as shown in Fig 1.1.3.

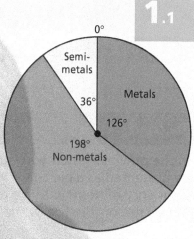

FIG 1.1.3 Pie chart showing the classification of the first 20 elements

Activity 1.1.1

Collecting data

Here is what you will need:

- Notepad and pencil
- Compass
- Protractor.

Here is what you should do:

1. Go outside into the school field or an undeveloped area near to where you live.

2. Look carefully at the colours of the flowers on each plant. Place each into one of the following groups: white, yellow, orange, red, green, blue, purple.

3. Record your data in a frequency table using a tally.

4. The number of plants in your sample should be a multiple of 18, e.g. 18, 36, 54 or 72, depending on how many plants are in flower at that time.

5. Present the data as a pie chart.

Check your understanding

1. Table 1.1.3 shows the number of animals from different groups seen on a piece of land.

Animal group	Frequency
Birds	8
Insects	16
Reptiles	5
Amphibians	3
Mammals	4

TABLE 1.1.3

a) How many animals were observed in total?

b) How many degrees should represent each animal in a pie chart?

c) Draw a pie chart to represent this information.

Key terms

data another word for information

frequency the number of items of a particular value

pie chart method of displaying data as sectors of a circle

Presenting data in different ways

Pie charts, tables and bar charts 》》

In the previous lesson you learned how data may sometimes be presented as a pie chart. In this form, the number or frequency of items within a group is reflected by the size of the **sector**, or the slice of the pie.

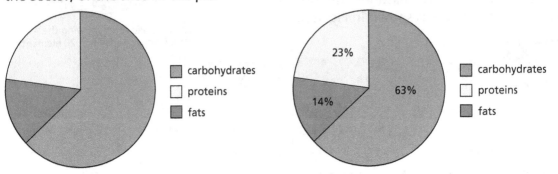

FIG 1.2.1 A person's daily diet shown as a pie chart

The pie chart above shows the proportion of food groups in a person's daily diet. The first chart gives no **quantitative** information. We can deduce only that carbohydrates form the majority of their diet, then proteins and finally fats.

We could use a protractor to measure the angles and from this, find the relative sizes of the sectors. This gives quantitative information which can be shown as a table. This information can be used to present the data in a different way, such as a bar chart (Fig 1.2.2).

Food group	Percentage in diet
Carbohydrates	63
Proteins	23
Fats	14

TABLE 1.2.1

Notice that the charts only show the proportions of the different food groups eaten. We need more information to find the actual amounts of the different food groups eaten.

If the total amount of food eaten in the day was 2.4 kg then we can calculate that the person ate:

• 2.4 × 0.63 = 1.512 kg of carbohydrates
• 2.4 × 0.23 = 0.552 kg of proteins
• 2.4 × 0.14 = 0.336 kg of fats

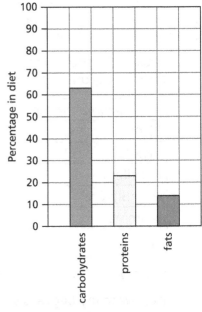

FIG 1.2.2 A person's daily diet by proportion shown as a bar chart

Pie charts and graphs

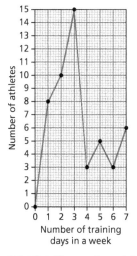

We would not normally draw a graph to represent three items of data, but when a pie chart shows a greater number of items, they can also be displayed as a graph.

The following pie chart shows how many days each week the athletes at a club train. This data can also be represented as a graph (Fig 1.2.4).

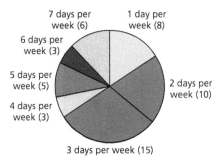

FIG 1.2.3 The number of days per week athletes train as a pie chart

FIG 1.2.4 The number of days per week athletes train as a line graph

---

Activity 1.2.1

Using the data in a pie chart to create a graph

The pie chart in Fig 1.2.5 contains information about the sources of energy used in a country.

1. Write the data given in the pie chart as a table.

2. Use the data in your table to draw a graph.

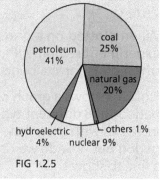

FIG 1.2.5

Check your understanding

1. The following bar chart contains data about the eye colour of students in a class. Represent the same data as a pie chart.

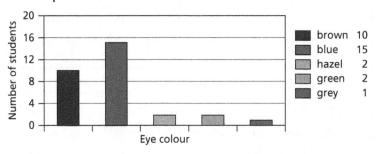

brown	10
blue	15
hazel	2
green	2
grey	1

FIG 1.2.6

Key terms

sector part of a circle enclosed by two radii and the arc between them

quantitative relating to numbers

Patterns and anomalies in data

We are learning how to:

• identify patterns in data
• draw a line of best fit on a graph
• identify readings in a set of data which are anomalous.

Patterns »

In science a set of data items often show a **pattern** or a **trend**. They are not a set of random values.

The pattern in Fig 1.3.1 shows that as the potential difference across a resistor increases, the current increases in the same proportion.

Patterns do not always produce straight-line graphs. The mass of products from a chemical reaction changes over time to produce a curved graph (Fig 1.3.2).

Line of best fit

When plotted on a grid, the data obtained from an experiment seldom all fit on a straight line or curve. More often plots form a band.

Fig 1.3.3 shows a graph of average life span against gestation period for 8 different types of animal. Average life span clearly increases with gestation period but how do we draw a line to show this?

Joining the points is not the answer since the short lines joining adjacent points will not always show the same trend. To show the pattern we need a single **line of best fit**. This may be a straight line or a curved line depending on the data.

Notice that a line of best fit might not pass through any of the points on a graph, but it will be the line that best represents the pattern or trend created by them.

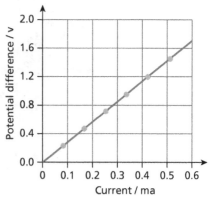

FIG 1.3.1 A linear pattern

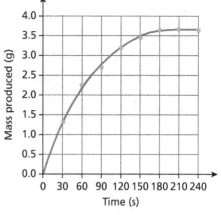

FIG 1.3.2 A non-linear pattern

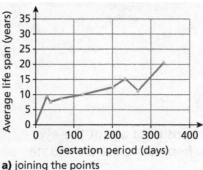

a) joining the points

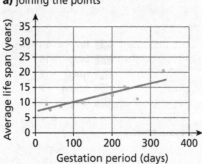

b) line of best fit
FIG 1.3.4 Line of best fit

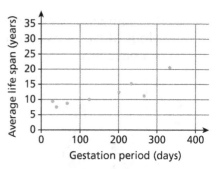

FIG 1.3.3 Average life span and gestation period

Anomalous values

An **anomalous** value is one which doesn't fit a set of data. Anomalous values are often the result of taking incorrect measurements during an experiment.

In Fig 1.3.5 the plot inside the red circle clearly doesn't fit the pattern formed by the remaining plots. It is important to identify an anomalous value and exclude it when determining the line of best fit.

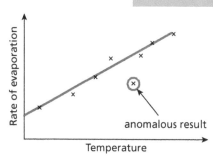

FIG 1.3.5 Anomalous result

Activity 1.3.1

Line of best fit

Here is what you need:

- Beaker 100 cm³
- Thermometer
- Hot water
- Stopwatch or clock.

1. Place the thermometer in the hot water and measure its initial temperature.

2. Measure the temperature of the water every minute for 15 minutes and record the values in a table.

3. Plot a graph of temperature against time for the water. Draw the line of best fit through the plots.

4. Suggest how an anomalous result might be obtained during this activity. Identify any plots from your data that you think are anomalous.

Check your understanding

1 **a)** Which of the plots in Fig 1.3.6 appears to be an anomalous value?

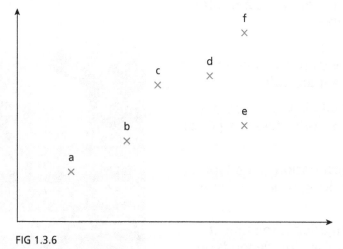

FIG 1.3.6

b) What account should be taken of the anomalous value when drawing the line of best fit?

Fun fact

Anomalous values are not always errors. Sir William Ramsay noticed that nitrogen obtained from air had a slightly different density to nitrogen made from ammonia. This anomaly led to the discovery of the gas argon.

Key terms

pattern regular sequence

trend pattern of gradual change

line of best fit line that best shows the trend exhibited by data

anomalous deviating from an expected pattern

Variables

We are learning how to:

• identify a variable
• use different types of variables in scientific experiments.

Constants and variables »

A **constant** is something which always has the same value. Numbers like 4, 7, 23, etc. are all constants. In mathematics you might also have come across the constant π, which has an approximate value of $\frac{22}{7}$ or 3.14.

A **variable** is something that can have different values. You come across many variables in everyday life but you probably don't think of them as such. Here are some examples:

- The number of students playing in the compound during break.

- The number of birds visiting a particular tree in a 5 minute period.

- The number of minutes it takes a student to complete his or her homework.

Variables are often represented by letters in mathematical equations like $y = x + 2$.

Variables in scientific experiments

In scientific experiments there are three types of variables:

- An **independent variable** is one that you change.

- A **dependent variable** is one that changes as a result of changing the independent variable.

- A **control variable** is one that could take lots of different values but you keep constant to make sure that the results of your experiment are valid.

Let's see how it fits into a simple activity in which you are going to investigate how long it takes for different types of sugar to dissolve in water.

The independent variable in this investigation is the type of sugar, for example granulated sugar, icing sugar, demerara sugar, etc.

The dependent variable in this investigation is the time it takes for each sample of sugar to dissolve. This depends on the type of sugar and is likely to be different for each sample.

In order to identify the control variables we need to think of things that might affect the amount of time needed for each sugar sample to dissolve. For example:

> **Fun fact**
>
> In experiments on living things a control is often used. The control is the same type of organism as the one used in the experiment but nothing is done to it. At the end of the experiment the organism can be compared to the control to makes sure that any changes are not the result of natural processes.

FIG 1.4.1 Dissolving sugar

- the mass of sugar
- the volume of the water
- the temperature of the water
- stirring – sugar dissolves faster if the water is stirred.

In order to make a fair test or a fair comparison between the different types of sugar all of the above must be controlled. This involves having a fixed value for each one.

Activity 1.4.1

Comparing rates of reaction

Calcium carbonate reacts with dilute hydrochloric acid to produce carbon dioxide gas. The volume of gas given off can be measured. Calcium carbonate can be obtained in different forms containing different-sized particles.

FIG 1.4.2 Different forms of calcium carbonate

Plan an experiment to compare how quickly different forms of calcium carbonate will react with dilute hydrochloric acid. To do this you must:

1. Identify the independent variable.
2. Identify the dependent variable.
3. Identify any other factors that might affect the validity of the results obtained and how these should be handled as control variables.

Check your understanding

1. Here is an experiment to test how well three samples of fibreglass (A, B, C) reduce heat loss.

   A layer of each type of fibreglass 1 cm thick was placed at the bottom and around the sides of three 250 cm³ beakers. A thermometer was placed upright in each beaker.

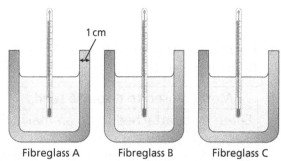

1 cm

Fibreglass A   Fibreglass B   Fibreglass C

FIG 1.4.3

   150 cm³ of water at a starting temperature was placed in each beaker. The temperature of the water in each beaker was recorded every 2 minutes for the following 30 minutes. Identify the following variables:

   a) independent   b) dependent   c) control.

Review of Working like a scientist

- Data is another word for information.

- Raw data is data as it is collected before it is organised in any way.

- Tally charts and frequency tables are used to organise data.

- Data can be displayed in a number of ways including a pie chart.

- The same data can be displayed in different ways.

- Data often shows a pattern or trend.

- Anomalous data values are those which don't appear to fit an obvious pattern or trend.

- A constant is something that always has the same value.

- A variable is something that can have different values.

- An independent variable is the one you change during an experiment.

- A dependent variable is one which changes as a result of changes to the independent variable in an experiment.

- Control variables are variables which are kept constant in order to ensure that the results of an experiment are valid and a fair test is carried out.

- Drawing scientific specimens requires a sharp pencil.

- Different techniques are used to draw natural objects including unbroken and broken lines, stippling, streaking and cross-hatching.

- Scientific drawings must be labelled and given a meaningful title.

Review questions on Working like a scientist

1. A survey of how the different flowering plants in a garden produce seeds gave the following results.

Method used to disperse seeds	Tally	Frequency
Eaten by or attached to an animal	IIII IIII II	
Carried by the wind	IIII IIII IIII	
Thrown by plant as fruit dries out	IIII IIII	
Carried by water	I	
	Total	

TABLE 1.RQ.1

a) Copy and complete Table 1.RQ.1

b) The data is to be displayed as a pie chart. How many degrees should represent each type of flowering plant?

c) Draw a pie chart to represent this data.

2. A student carried out an experiment to determine the conditions most favoured by woodlice.

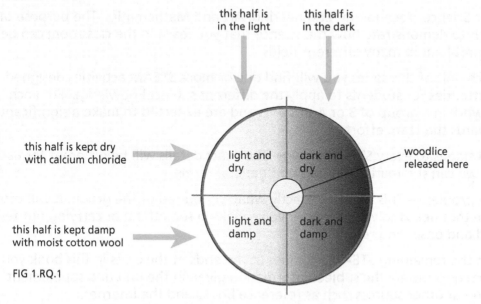

this half is in the light

this half is in the dark

this half is kept dry with calcium chloride

light and dry

dark and dry

woodlice released here

this half is kept damp with moist cotton wool

light and damp

dark and damp

FIG 1.RQ.1

A group of 20 woodlice were released at the centre of a circular container. The conditions in each quarter of the container were different. After 2 minutes the number of woodlice in each quarter was counted.

a) Give the independent variable.

b) Give the dependent variable.

c) Give two control variables.

3. A student has drawn and labelled a diagram of a cell. Unfortunately she did not follow all of the labelling guidelines used in science. Identify as many things as you can that show she did not follow the guidelines.

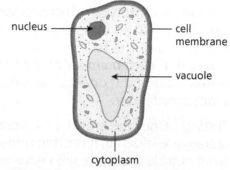

nucleus

cell membrane

vacuole

cytoplasm

FIG 1.RQ.2

4. The photograph on the right shows a mixture of solid and liquid being poured from a beaker into a filter paper in a filter funnel. The liquid is being collected in a conical flask which is supported by a stand and clamp.

Draw a labelled diagram of this apparatus.

FIG 1.RQ.3

What are STEAM activities?

STEAM stands for **S**cience, **T**echnology, **E**ngineering, **A**rt and **M**athematics. The purpose of STEAM activities is to demonstrate how the science that you learn in the classroom can be applied to solve problems in many different fields.

At the end of each unit of this series you will find one or more STEAM activities designed to provide opportunities for students to apply the different subject knowledge. For each activity you will work in a group of 3 or 4 students and are expected to make a significant contribution towards the team effort.

There are several stages to every STEAM activity. The exact details will be different for each specific task, but we can summarise the stages in general terms.

1. Defining the problem – This tells you exactly what is required of the group. It will often be a problem that needs solving. It may involve making something or carrying out tests of some kind and observing results.

2. Research – In the remaining STEAM activities at the ends of the units in this book you will be expected to review the subject information given in the unit and supplement this by looking at other sources such as reference books and the internet.

3. Requirements – You need to think about what you will need and make lists. You may need a list for materials and another list for tools and equipment. You might also need to obtain advice from someone outside your group, such as your teacher.

4. Agree on a solution – You are working as one student in a team of 3 or 4. The members of the team will probably suggest more than one solution to the problem. If so, then team members need to analyse and evaluate the suggestions of the different members and come to an agreed solution.

5. Building a prototype – This provides the team members with an opportunity to show their manual skills. The task should be organised in such a way that each member of the team contributes. It may be useful if one person takes on the role of team coordinator.

FIG 1.SIP.1 Testing egg parachutes

6. Testing – Once the prototype is built it needs to be tested or evaluated. It is often useful to have people from outside the team, who have not been closely involved in the planning and building, to review your work. A fresh pair of eyes can sometimes pick up things missed by people who are closely involved. Consider the comments made by your reviewers and modify what you have made in the light of them if you think it is appropriate.

7. Reporting back – The final task is to compile and to deliver a report of some kind to explain what you did and to show your end product. This can take various forms, such as a PowerPoint presentation or a demonstration of what you have designed and built.

If possible, take pictures at different stages of construction and testing on your cell phone. Illustrated reports are always more interesting than just writing and description.

Making a hair hygrometer >>>

1. A hygrometer is a device that measures humidity, which is the amount of water vapour in the air. A human hair increases in length slightly according to how humid the air is. Your task is to make a simple hair hygrometer.

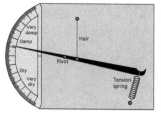

FIG 1.SIP.2 Ideas for a hair hygrometer

2. The first hair hygrometer was made by the Swiss scientist Horace Benedict de Saussure in 1783. Carry out some research in available books or by keying 'hair hygrometer' into a search engine on the internet.

3. The longer the hair, the greater the change in length due to humidity. To build a hair hygrometer, you will need to find someone with long hair who is prepared to donate one in the interests of science.

   The next problem is how to translate the change in length to a reading on a scale. Here are some designs that might give you some ideas.

4. The team needs to decide on a design. In your discussions you should consider such factors as availability of materials, ease of construction and durability when used. Don't opt for a difficult design and construction because you think it looks 'more scientific'. The easiest solutions to problems are often the best.

5. Make a list of material and tools you are going to need to build your prototype. Make use of readily available materials if you can. For example, a wooden base looks good but a piece of thick cardboard from a carton will work equally well. If you haven't got a glue gun, use spots of ordinary glue and allow them to harden.

   Place your instrument in a humid atmosphere and in a dry atmosphere so that you can calibrate the scale. A humid atmosphere might be near a boiling kettle, whereas a dry atmosphere might be achieved using a hair dryer on a low setting. You will only be able to create a simple qualitative scale from 'dry' to 'wet'.

   If possible, take some photographs during the construction and testing. These will be useful for illustrating your report.

6. Once your instrument is complete it needs to be tested.

   You could compare your hair hygrometer with an actual analogue or digital hygrometer if one is available in the science department.

FIG 1.SIP.3 Hygrometers

7. Your final task is to demonstrate your hair hygrometer and to discuss its construction. Use photographs to illustrate your discussion.

Unit 2: Photosynthesis and energy relationships

We are learning how to:

- investigate the requirements for photosynthesis
- understand how leaves are adapted for photosynthesis
- describe food chains and food webs
- appreciate the impact of humans on feeding relationships.

Photosynthesis and energy relationships »

We associate the colour green with fields and trees, and anywhere else where there are plants. The reason for this is that all plants contain a green pigment called **chlorophyll**. The importance of chlorophyll lies not in its colour but its ability to trap energy from sunlight, which plants use to make food.

FIG 2.1.1 Photosynthesis is the process by which green plants make their own food

> **Fun fact**
>
> The process of photosynthesis is not very efficient. Typically, only around 5% of the energy shining on a plant is used to form biomass.

The process by which plants absorb energy from sunlight to make food is called **photosynthesis**. It can be represented by the following word equation:

carbon dioxide + water + energy ⟶ glucose + oxygen

The leaf cells absorb carbon dioxide from the atmosphere and water from the ground through the plant roots. Oxygen is released to the atmosphere and the glucose is stored in the leaf as starch until it is needed elsewhere by the plant.

Nutrition in green plants is very different to that in animals. In green plants the small molecules carbon dioxide and water are combined to form larger molecules in the

5. Place the plant in a bright sunny position for two to three hours.

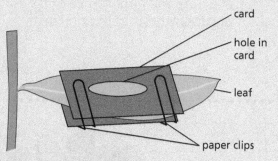

FIG 2.3.2

6. Remove the leaf from between the cards and test it for starch using iodine solution. If starch is present in any part of the leaf it will turn blue-black.

7. Which part of the leaf gave a positive test and which gave a negative test?

FIG 2.3.3 Variegated leaves

Some plants have leaves that are not completely green. There are areas of the leaf that do not contain the green pigment chlorophyll. These are called variegated leaves.

Variegated leaves can be used to show that chlorophyll is necessary for photosynthesis.

When a variegated leaf is tested for the presence of starch only those parts that are green will turn blue-black. No starch is present in those parts of the leaf that do not have chlorophyll.

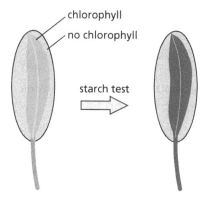

FIG 2.3.4 Starch test on a variegated leaf

Fun fact

The leaves of all green plants contain chlorophyll.

FIG 2.3.5 Sometimes chlorophyll is masked by other pigments

Check your understanding

1. Fig 2.3.6 shows a plant growing under a bell jar. The dish contains a substance called soda lime, which absorbs carbon dioxide from the air.

   How could you use this plant and apparatus to demonstrate that photosynthesis requires carbon dioxide?

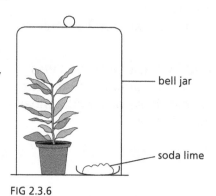

FIG 2.3.6

Key terms

ecosystem community of interacting organisms and their physical environment

photosynthesis the process by which green plants make their own food

The importance of the products of photosynthesis

We are learning how to:

• evaluate the importance of photosynthesis to life on Earth.

Glucose and starch >>>

In the previous lesson you learned that, as a result of photosynthesis, green plants are able to make glucose. Glucose is of fundamental importance as a source of energy. It is broken down in both plant and animal cells to release the energy. This process is called **cell respiration**. The energy released is needed to drive the chemical processes within a living organism that maintain life.

Plants therefore use up some of the glucose they produce and store what remains as starch. In some plants the store of starch can be extremely large.

Animals obtain food from eating all parts of plants, their fruit and their seeds.

Green plants are producers of food. A **food chain** always starts with a green plant or something produced by it. If there were no green plants there would be no food chains and life would not exist in its present form on Earth.

Oxygen

The other product of photosynthesis is the gas **oxygen**. This is equally important in maintaining life on Earth. Oxygen makes up around 21% of the gases in air.

The following processes require oxygen from the air:

• It is used up by all living things during cell respiration.
• It is used up during the burning of fuels.
• It is used up when metals corrode, like iron becoming rusty.

If the amount of oxygen removed by these processes was not continually **replenished**, the Earth would have run out of oxygen a long time ago. Photosynthesis provides a means of maintaining the level of oxygen in the air.

FIG 2.4.1 Root vegetables are stores of starch

Fun fact

Air contains about one-fifth oxygen so it can be thought of as 'dilute' oxygen. Chemical reactions like burning are far more vigorous in pure oxygen than they are in air.

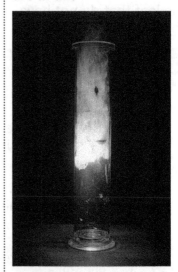

FIG 2.4.2 Magnesium burning in pure oxygen

Obtaining oxygen from photosynthesis and testing it

Here is what you need:

- Pondweed (*Elodea*)
- Test tube
- Teaspoon
- Beaker
- Distilled water
- Wooden splint
- Glass funnel
- Sodium hydrogencarbonate
- Flame.

Here is what you should do:

1. Three-quarter fill a beaker with distilled water and add a teaspoonful of sodium hydrogencarbonate. Stir the water until the solid dissolves. This will ensure that there is plenty of dissolved carbon dioxide in the water.

2. Place some freshly cut pondweed in the bottom of the beaker and cover it with an inverted glass funnel.

3. Fill a test tube with distilled water. Cover the end and invert the test tube so it sits on the inverted funnel.

4. Place the apparatus in a sunny area.

5. Pondweed can take a while to start photosynthesising so leave the apparatus for several hours or overnight to gather a test tube full of oxygen.

6. Keeping your thumb over the test tube, turn the tube the right way up.

7. Light a wooden splint and then blow it out so that the end is smouldering and glowing red.

8. Introduce the glowing wooden splint into the top of the test tube.

Objects burn more vigorously in pure oxygen than they do in air. Oxygen will cause a glowing wooden splint to relight. This is a simple test for this gas.

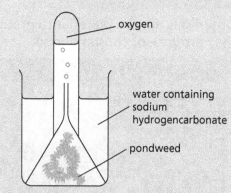

FIG 2.4.3

oxygen

water containing sodium hydrogencarbonate

pondweed

Check your understanding

1. Which of the following processes require oxygen?

   burning   cell respiration

   photosynthesis   rusting

2. Explain why a glowing wooden splint relights in pure oxygen but not in air.

3. What is obtained as a result of cell respiration and how is it used?

Key terms

cell respiration process by which energy is released in the cells of living organisms

food chain feeding relationship between organisms

oxygen gas which makes up about one-fifth of air and is essential for life

replenish to replace

Producers, consumers and decomposers

We are learning how to:

- identify producers, consumers and decomposers.

Producers and consumers

Organisms may be placed into groups according to whether they produce or consume food.

FIG 2.5.1 Green plants are called producers because they use energy from sunlight to produce the food

Green plants use some of the food themselves and what remains is often eaten by animals.

FIG 2.5.2 Animals are called consumers because they consume the food made by green plants

Key terms

primary consumers consume food from green plants directly

secondary consumers eat the animals that eat only plants

Animals that eat only plants are called herbivores. Herbivores such as cattle are **primary consumers** because they eat plants directly.

Animals that eat only other animals are called carnivores. Carnivores like the leopard are **secondary consumers**. They do not eat plants directly but eat herbivores that themselves eat plants.

Decomposers

There is another important group of organisms, which feed on dead organic material. This might be dead plants and animals or animal waste. This group is called the decomposers.

Some decomposers, such as bacteria, are so small that they can only be seen with the help of a microscope. Fungi are decomposers. They often grow on decaying plant material or on animal waste.

FIG 2.5.3 Carnivores such as leopards are secondary consumers

Although decomposing organic material does not look or smell very nice, decomposers do an extremely important job. By breaking down the waste, nutrients are released into the soil that can be used again by other organisms.

Activity 2.5.1

Identifying producers and consumers

Your teacher will lead this activity.

Here is what you need:

- Notebook.

Here is what you should do:

1. Take a short walk around the school compound or the grounds nearby.

2. Make a note of any organisms you see and decide whether they are producers, consumers or decomposers.

3. Make lists of the two groups of organisms.

FIG 2.5.4 Worms are decomposers: as they feed, they move organic waste through the soil

Check your understanding

1. Maize grows in a field. The maize plants are eaten by locusts. The locusts are eaten by lizards.

   a) State whether each of the organisms described is a primary consumer, a secondary consumer or a producer.

   b) How are the nutrients contained in the bodies of these animals ultimately returned to the ground?

> **Fun fact**
>
> Vultures are scavengers.
>
>
>
> FIG 2.5.5 Scavengers are secondary consumers that do not kill primary consumers but feed on their dead bodies

Food chains

We are learning how to:

• draw and interpret a food chain.

Food chains »

Organisms can be grouped as producers or consumers depending on whether they make food or eat food. The relationship between the two groups can be shown as:

producers ⟶ consumers

The arrow joining the two groups of organisms can be interpreted in several ways.

• It can be taken to mean 'is eaten by' – producers are eaten by consumers.

• It shows the transfer of energy – consumers obtain energy by eating producers.

• It shows the transfer of nutrients – consumers obtain nutrients by eating producers.

If you show this relationship using the names of organisms, the result is called a **food chain**.

Grass ⟶ Rabbit

You can add this to the food chain.

Grass ⟶ Rabbit ⟶ Ocelot

At each link in the food chain, energy and nutrients pass from one organism to the next.

Plant ⟶ Insect ⟶ Lizard ⟶ Snake ⟶ Hawk

There are two important facts that are true of all food chains.

1. A food chain always starts with a producer – a green plant or something formed from it.

2. All of the energy that passes along a food chain comes from the Sun.

FIG 2.6.1 A rabbit is a herbivore and therefore a primary consumer

FIG 2.6.2 A rabbit may be eaten by a carnivore (secondary consumer) such as an ocelot

Activity 2.6.1

Making food chains using name cards and arrow cards

Here is what you need:

- 12 blank cards (4 cm × 3 cm).

Here is what you should do:

1. Make name cards for each of these organisms: Barbados cherry tree, stink bug, praying mantis, butterfly, caterpillars, slugs, toads, small birds, hawks.

2. Make three arrow cards.

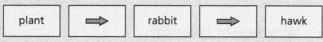

FIG 2.6.3 An example of a food chain

3. Use your cards to make some food chains. Each of your food chains:

   - should start with a plant
   - should have at least three organisms.

4. Write down your food chains.

Key term

food chain relationship between producers and consumers using the names of organisms

The general flow diagram for a food chain is:

Producer ⟶ Primary consumer ⟶ Secondary consumer

This is an example of a food chain.

Grass ⟶ Grasshopper ⟶ Lizard ⟶ Snake ⟶ Falcon

Check your understanding

1. Fig 2.6.4 shows a food chain from a lake in Europe.

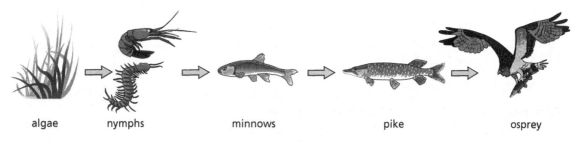

FIG 2.6.4

a) Which organism feeds on
   i) algae?
   ii) nymphs?
   iii) minnows?
   iv) pike?

b) From the organisms in the food chain, name:
   i) a primary consumer
   ii) a secondary consumer
   iii) a producer.

Food webs

We are learning how to:

• draw and interpret a food web
• predict how changes to the population of one organism in a food chain or web will affect the populations of other organisms.

Food webs 〉〉〉

A food chain shows each animal eating one other type of organism. In the real world, animals eat lots of different things. Each animal, therefore, will appear in many food chains.

Here are some food chains involving organisms that live in a desert.

Desert plants ⟶ Insects ⟶ Large lizards ⟶ Hawks

Desert plants ⟶ Desert rats ⟶ Snakes ⟶ Hawks

Most of the organisms are in more than one food chain. The feeding relationships between these organisms are better shown as a **food web**.

A food web shows how food chains are linked together.

If we were to take a food chain in isolation, we could easily predict the effect that a fall in the population of one organism would have on the others. For example, suppose that as a result of pollution from a factory, all of the large lizards in the following food chain died.

Desert plants ⟶ Insect ⟶ Large lizards ⟶ Hawks

We could predict that the populations of insects would increase, because the large lizards were no longer eating them, while the hawks would all die out, as they would have no large lizards to eat.

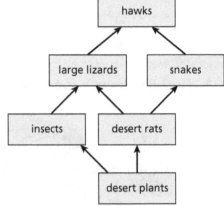

FIG 2.7.1 Food web

Life however, is not so simple:

• Organisms at the bottom of a food chain are generally eaten by more than one type of **consumer**.

• Organisms at the top of a food chain generally eat more than one type of **prey**.

Here are some predictions we might make from the position of large lizards in the food web:

• A rise in the numbers of insects and small lizards – as large lizards will eat fewer of them.

• A fall in the numbers of desert plants – because there will be more insects and small lizards feeding on them.

6. Fig 2.RQ.5 shows the inside of a cassava root.

FIG 2.RQ.5

a) What substance forms most of the cassava root?

b) Describe a simple chemical test that can be used to confirm the identity of this substance.

c) How is this substance formed in the cassava plant?

d) What evidence is there that the cassava plant produces far more of this substance than it needs to grow?

e) Where would a cassava root be placed in a food chain?

7. A biologist investigated feeding relationships around a large oak tree near where she lives. Here are some of the observations from her field book.

Beetles are eating the bark of the oak tree.

A hawk caught and ate a woodpecker.

Blue tits are eating caterpillars.

Caterpillars are eating the leaves on the tree.

A woodpecker is eating beetles on the ground.

A blue tit is caught and eaten by a hawk.

Two woodpeckers are eating caterpillars.

An owl caught and ate a blue tit.

a) Draw a food web using these observations.

b) From your food web give one example of:

   i)   a producer

   ii) a primary consumer

   iii) a secondary consumer.

c) A disease killed off the woodpecker population in and around the tree. Make two predictions about how this will affect the other populations of organisms in the food web. You should explain each prediction.

Photosynthesis and energy relationships

Animals seldom have only one source of food. They may eat different foods according to season and/or availability. It is for this reason that an animal may feature on several different food chains in an ecosystem. Food chains can be combined to form a food web. A food web gives a more complete picture of the feeding relationships between organisms in an ecosystem.

Identify a small piece of land either within your school compound or close to your school or home. This land

FIG 2.SIP.1 A small piece of land

should be undeveloped and contain different types of plants.

You may assume that there are plans to build a path on this land and install a bench. This will allow students or local people to sit and enjoy the surroundings.

The Nature Conservancy of Jamaica is concerned that this work will drastically affect the organisms that live on and near this land. They have asked you to construct a food web involving these organisms in order to better assess the impact of any changes. They are hoping that your report will allow them to identify the position for the path and bench that will cause least disturbance.

1. You are going to work in groups of 3 or 4 to investigate the feeding relationships between organisms on this land. The tasks are:

   - To review the significance of food chains and food webs and how to construct them.
   - To carry out research into some of the plants and animals you can expect to find in and around the land so that you will be able to identify them. You will also need to know something about the feeding habits of the animals so you can identify likely food sources.
   - To spend time observing the organisms on and around the land and gathering information about their feeding habits.
   - To construct a provisional food web for the land.
   - To carry out further observations on the land to determine how accurate your food web is, and to supplement it with further observations.
   - To compile a report, including a PowerPoint presentation, describing how you gathered the information needed to construct a food web. You should be prepared

to recommend likely positions for a path and bench and justify them in terms of least impact. You should illustrate your report as much as possible with photographs, a simple map and any other relevant diagrams.

a) Look back through the later lessons in this unit and make sure you understand the general structure of a food chain, and how two or more food chains may be combined to form a food web.

b) Carry out some research into common wild flowers and other plants you are likely to find.

FIG 2.SIP.2 Some common wild plants

You cannot hope to include all of the wild plants found in Jamaica so focus on those which are common and well known. Find out what you can about the animals that live near or feed on these plants. For example, are hibiscus pollinated by bees? What birds eat wild raspberries?

c) Take an overview of the land, making approximate measurements of its length and width and use these to draw a simple map. This will provide a framework within which you can work. Carry out your first observations of the plants and animals in different places on the land and what appears to eat what. Take pictures of the organisms that will feature in your food web.

d) Draw up food chains based on your observations and then use them to create a provisional food web. This will give you a better overall picture of the relationship between organisms and help you to focus on likely areas for your future observations.

e) Carry out further observations of the plants and animals on and around the land. Check that your provisional food web is correct as far as it goes and be prepared to correct any mistakes. Gather more information that can be used to give a clearer and more precise food web. Take pictures of any new organisms that will be added to your food web.

f) Write a report summarising your findings and use this as the basis of a PowerPoint presentation. Your report, and presentation, should contain:

- A statement about why the work was undertaken.
- A brief account of the land with a simple map.
- An illustrated account of the evidence you have been able to gather in regard to feeding relationships.
- A food web for the piece of land with as much detail as you can.
- A recommendation for where a path and bench should be located based on your judgement of where they will cause least disruption.

Unit 3: More about matter

We are learning how to:

- distinguish metals and non-metals
- appreciate how the elements are grouped in the Periodic Table
- use the Periodic Table to predict atomic structure.

More about matter ≫

Elements

Elements are substances that cannot be broken down into anything simpler. There are 94 elements that occur naturally on the Earth although some of them exist only in very small quantities. There are also a few elements which have been made artificially using high-energy devices called particle accelerators.

FIG 3.1.1 **a)** Metallic elements   **b)** Non-metallic elements

Elements like copper and aluminium are metals, whereas elements like sulfur and carbon are non-metals.

Under normal conditions of temperature and pressure, most elements are solids, however, some are gases and a very small number are liquids. All elements can be solid, liquid or gas given appropriate conditions.

Atoms

An atom is the smallest particle of an element that can take part in a chemical reaction. These are the building blocks from which all substances are formed. All of the atoms of an element are similar to each other, but are different from the atoms of another element.

hydrogen   oxygen   carbon

FIG 3.1.2 Atoms of different elements

- Some metals can be polished to a high lustre.

Properties of non-metals

Non-metals are far more diverse in their appearance.

Non-metals share some common properties.

- Non-metals are often soft and have low melting points and boiling points. A number of non-metals are gases at room temperature.

- All non-metals are poor conductors of heat and, with the exception of a form of carbon called graphite, they are poor conductors of electricity.

Activity 3.3.1

Examining metals and non-metals

Here is what you need:
- Samples or pictures of some different elements

Here is what you should do:

1. Look at each element carefully.

2. Make a list of their features and determine whether the element is a metal or a non-metal.

Properties of semi-metals

At first glance you might think that silicon is a metal but it is actually a semi-metal. It has a mixture of properties.

Metallic properties:

- It is hard and shiny, and has a high melting point and boiling point.

Non-metallic properties:

- It is a poor conductor of heat and a much weaker conductor of electricity than a metal. It is sometimes described as a **semi-conductor**.

Check your understanding

1. Use the Periodic Table to determine which of the following elements are metals, and which are non-metals. Make two lists.

oxygen	sulfur	iodine	argon
magnesium	copper	silver	calcium
chlorine	hydrogen	iron	sodium
nitrogen	mercury	zinc	
tin	phosphorus	gold	

FIG 3.3.2 Metals

FIG 3.3.3 Non-metals

FIG 3.3.4 Silicon is a semi-metal

Key terms

semi-metal substance that has some metallic properties and some non-metallic properties

conductor of heat allows heat energy to pass through it

conductor of electricity allows an electric current to pass through it

semi-conductor substance that conducts electricity but not as well as a metal

Groups of the Periodic Table

We are learning how to:

- identify different groups in the Periodic Table
- briefly describe elements in some groups.

The vertical columns of the Periodic Table are called groups. There are similarities in the chemistry of all the elements in any group. You will see from the table in lesson 3.2 that the groups are numbered from 1 to 18. In this lesson we will look at elements in some of these groups.

Group 1 ≫

The '**alkali metals**' is the traditional name of the Group 1 metals. These are soft metals, easily cut with a knife. They are light grey in colour and form ionic compounds which are white or colourless.

FIG 3.4.1 Group 1 elements used in the laboratory

The Group 1 metals all readily react with air, water and dilute acid and are stored in oil. Reactivity increases passing down the group so lithium is the least reactive element in the group. The metals beyond potassium are too reactive for safe use in the laboratory and therefore the chemistry of this group is usually limited to the first three elements.

Notice that hydrogen is often shown at the top of Group 1 in the Periodic Table simply because of the structure of the hydrogen atom. Hydrogen is not a Group 1 element.

Activity 3.4.1

Lithium reactions

Your teacher will demonstrate this activity because the reaction of the Group 1 metals with water is potentially hazardous.

Very small pieces of the metals should be used. The reactions should be carried out behind a safety screen.

Here is what you will need:

- Beaker 500 cm³ x 3
- Knife
- Distilled water
- Lithium
- Potassium
- Sodium
- Universal Indicator.

1. Cut a tiny piece of lithium, no bigger than grain of rice
2. Three-quarter fill a beaker with water.
3. Place the lithium onto the surface of the water.
4. Observe until the reaction is complete.
5. Place a few drops of universal indictor into the water.
6. Repeat steps 1–5 with pieces of sodium and potassium.
7. Describe what happens when lithium, sodium and potassium react with water.
8. Place the three metals in order of reactivity, starting with the most reactive, on the basis of your observations.

Groups 3–12

Groups 3–12 of the Periodic Table are commonly called the **transition elements** or transition metals.

The transition metals share some similarities with the Group 1 metals, such as being good conductors of heat and electricity, but in other ways they are very different. The transition metals are generally hard and have high melting points and boiling points.

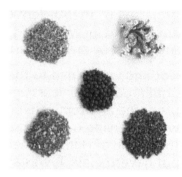

FIG 3.4.2 Commonly encountered transition elements: chromium, zinc, manganese, copper and iron

Group 17

The Group 17 elements are sometimes called the **halogens**.

Element	Symbol	Atomic number	Colour	State at rtp*
Fluorine	F	9	pale green	gas
Chlorine	Cl	17	yellow-green	gas
Bromine	Br	35	red-brown	liquid
Iodine	I	53	black	solid

TABLE 3.4.1   (* = room temperature and pressure)

These elements are all reactive, although reactivity decreases passing down the group. Fluorine is far too reactive to use safely in the laboratory.

FIG 3.4.3 Three common Group 17 halogens: chlorine (pale green), bromine (red brown) and iodine (pale purple)

Check your understanding

1. Rubidium is below potassium in Group 1. State what you would expect to happen, if a tiny piece of rubidium is placed in a trough of water containing universal indicator.

Key terms

alkali metals Group 1 elements

transition elements / metals elements of Groups 3–12

halogens Group 17 elements

Structure of an atom

We are learning how to:

- describe the structure of an atom
- compare the relative masses and charges on sub-atomic particles.

Concept of an atom ⟩⟩⟩

An atom is the smallest part of an element that can take part in a chemical reaction.

The idea of small units of matter, called atoms, was first proposed by the Ancient Greeks over 2 000 years ago. Their proposal was based on reasoning alone and was not supported by any scientific evidence.

Nothing was added to the understanding of the atom until around 200 years ago when the chemist John Dalton proposed an atomic theory to account for his observations when carrying out chemical experiments. We now know that some of Dalton's theory was not entirely correct but nevertheless, it was a major contribution to the understanding of the nature of atoms.

Atoms are too small to see so we need models and diagrams to visualise their structure.

Sub-atomic particles

Although atoms were originally regarded as indivisible, it became apparent to scientists working in the early part of the 20th century that an atom was itself made up of even smaller particles. These particles are described as **sub-atomic particles**.

An atom is composed of three types of sub-atomic particle: **protons**, **neutrons** and **electrons**.

At the centre of every atom is a nucleus. The nucleus is almost always composed of protons and neutrons. The only exception is the hydrogen atom, which normally contains only a proton.

Surrounding the nucleus are shells or orbits containing electrons. There is a maximum number of electrons that can exist in each shell. In the first shell the maximum is two, and in the second and third shells the maximum is eight.

The total number of electrons surrounding the nucleus of an atom is always equal to the number of protons in the nucleus. The number of neutrons in the nucleus of an atom is similar to but not always equal to the number of protons.

FIG 3.5.1 John Dalton lived from 1766 until 1844

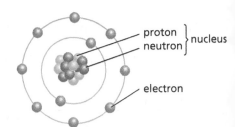

FIG 3.5.2 The sub-atomic particles in an atom

Fun fact

The **atomic number** is the number of protons in the nucleus of an atom.

The **mass number** is the total number of protons and neutrons in the nucleus of an atom.

The relative masses and charges on the sub-atomic particles are given in Table 3.5.1.

Particle	Relative mass	Relative charge	Position in the atom
proton	1	+1	In the nucleus
neutron	1	0	
electron	0	−1	Surrounding the nucleus

TABLE 3.5.1 Relative masses and charges of sub-atomic particles

You will observe from the information in the table that:

- the masses of a proton and a neutron are the same
- the mass of an electron is actually $\frac{1}{1836}$ th that of a proton and a neutron. This is so small that the value is usually taken as zero
- although the mass of a proton and an electron are very different, the charges they carry are equal in magnitude but opposite in value
- most of the mass in an atom is contained in the nucleus

It is possible to make models of atoms using modelling clay.

Key terms

sub-atomic particles protons, neutrons and electrons which form atoms

proton positively charged particle in the nucleus of an atom

neutron neutral particle in the nucleus of an atom

electron negatively charged particle surrounding the nucleus of an atom

Activity 3.5.1

Building a model of a lithium atom

The nucleus of a lithium atom contains three protons and four neutrons.

Here is what you will need:

- Items that represent sub-atomic particles such as playdough balls
- String   • Glue.
1. Decide what colours you are going to use to represent protons, neutrons and electrons.
2. Make an appropriate number of balls in each colour.
3. Place the protons and neutrons in a cluster at the centre to represent the nucleus of an atom.
4. Use string for the electron shells around the nucleus.
5. Place the electrons around the nucleus in appropriate shells.
6. Make a drawing of your atom in your note book.

Check your understanding

1. This question is about sub-atomic particles.

   a) Which particles are found in the nucleus of an atom?

   b) Which particles are charged?

   c) Which particles have a relative mass of 1?

   d) Which particles are arranged in shells around the nucleus of an atom?

Calculating numbers of sub-atomic particles

We are learning how to:

- describe the structures of atoms of different elements
- identify an element by the composition of its atoms.

Atoms of an element are characterised by their **atomic number** and their **mass number**.

- Atomic number = number of protons in the nucleus of the atom

- Mass number = total number of protons + neutrons in the nucleus of the atom

Fig 3.6.1 represents an atom of the element carbon. In the nucleus of this atom there are 6 protons and 6 neutrons therefore the atomic number of this atom = 6 and the mass number = 6 + 6 = 12.

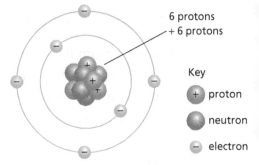

6 protons + 6 protons

Key

(+) proton

neutron

(−) electron

FIG 3.6.1 An atom of carbon

Atomic number and the Periodic Table ⟩⟩

If you look back at the Periodic Table in lesson 3.2 you will see that each element has a unique atomic number. Starting at the top left of the table and reading across:

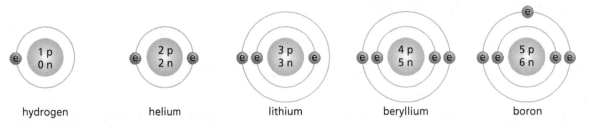

| hydrogen | helium | lithium | beryllium | boron |

FIG 3.6.2 Structures of the first five elements

If you study Fig 3.6.2 carefully you will see that:

- The atomic number = the number of protons and this increases by one each time.
- The number of electrons is always equal to the number of protons.

This means that we can predict the number of protons and electrons in an element from its atomic number. For example, the atomic number of sulfur is 16 therefore in an atom of sulfur there are 16 protons in the nucleus and 16 electrons surrounding it.

It also means we can identify an element from the number of protons contained in the nucleus of the atom. For example, an atom has 9 protons in the nucleus of its atom. The element which has an atomic number of 9 is fluorine.

Mass number and isotopes

Although all atoms of the same element always contain the same number of protons, these atoms may not always contain the same number of neutrons.

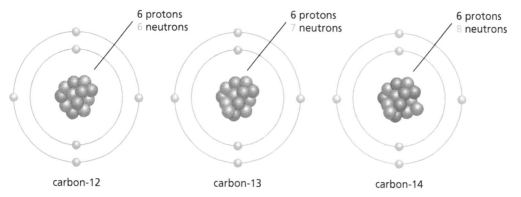

| 6 protons
6 neutrons | 6 protons
7 neutrons | 6 protons
8 neutrons |
| carbon-12 | carbon-13 | carbon-14 |

FIG 3.6.3 Three isotopes of carbon

Carbon atoms always contain 6 protons but may have 6, 7 or 8 neutrons. These different forms of carbon are called **isotopes** and are identified by their different mass numbers. Isotopes of an element seldom occur in equal amounts. The isotope carbon-12 is by far the most common isotope form carbon.

Activity 3.6.1

Making isotopes of hydrogen

There are three isotopes of hydrogen. Their structures are given in the following table.

Isotope	Number of protons	Number of neutrons
hydrogen-1	1	0
hydrogen-2	1	1
hydrogen-3	1	2

TABLE 3.6.1

Here is what you will need:

- Items to represent sub-atomic particles such as playdough balls
- Glue   • String.

1. Decide what colour you will use to represent each sub-atomic particle.

2. Make the number of each particle that you will need.

3. Construct an atom of each isotope and label it.

Check your understanding

1. State the number of protons and the number of electrons in an atom of the following elements.

   a) Oxygen   **b)** Aluminium   **c)** Lithium   **d)** Bromine

2. Name the element which has each of the following numbers of protons in its atom.

   a) 12   **b)** 7   **c)** 10   **d)** 20

Fun fact

All isotopes of the same element have exactly the same chemistry but may have slightly different physical properties.

Key terms

atomic number number of protons in the nucleus of an atom

mass number total number of protons + neutrons in an atom

isotopes atoms of the same element which have different numbers of neutrons

Diffusion

We are learning how to:

- describe how substances spread out by diffusion.

Diffusion >>>

We know from the kinetic theory that the particles in a gas move very quickly and frequently collide with other particles. No **kinetic energy** is lost during the collisions so the particles don't slow down.

The result is that particles quickly move in all directions. We describe this as **random motion**.

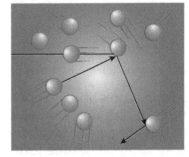

FIG 3.7.1 Random motion of particles

Activity 3.7.1

Determining the rate of diffusion

Your teacher will help you with this activity.

Here is what you need:

- Shallow open container like a watch glass
- Volatile liquid that has a characteristic smell, like a perfume
- Stopwatch with a second hand.

Here is what you should do:

1. Students sit or stand in rows at different distances from the teacher's table.

2. The teacher will pour a small amount of perfume onto a watch glass on the table at the front of the class.

3. Each student should record their distance from the perfume in the watch glass.

4. Each student should record the time taken for the smell of the perfume to reach them.

5. Tabulate the distances and times and calculate the rate of diffusion of the smell of the perfume.

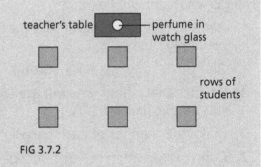

FIG 3.7.2

In Activity 3.7.1 the perfume spreads out, or diffuses, from the watch glass to all parts of the classroom. At the start the perfume is in a high **concentration** immediately above the watch glass and in a low concentration in the rest of the classroom. The perfume particles move from the region of high concentration to the region of low concentration.

We say that the perfume particles diffuse down a **concentration gradient**. A concentration gradient is the difference in the concentration of particles between two regions. After a while the concentration of perfume particles is the same in all parts of the classroom.

Diffusion also occurs in liquids and very slowly in solids.

Rate of diffusion

When ammonia gas and hydrogen chloride gas react they produce a white 'smoke' consisting of fine particles of ammonium chloride:

ammonia + hydrogen chloride ⟶ ammonium chloride

Ammonia gas is given off cotton wool soaked in ammonia and hydrogen chloride gas is given off cotton wool soaked in hydrochloric acid.

Look carefully at the position of the ring of white 'smoke' in Fig 3.7.3.

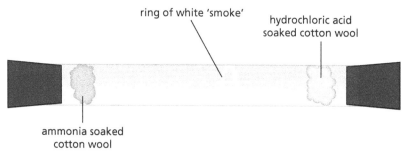

FIG 3.7.3 Diffusion of ammonia and hydrogen chloride

If ammonia gas and hydrogen chloride gas were diffusing at the same speed we would expect them to meet at the centre of the tube but the ring of white smoke forms nearer to the source of hydrogen chloride. This indicates that the ammonia gas particles are diffusing more quickly than the hydrogen chloride particles. This is evidence that substances do not diffuse at the same rate.

Check your understanding

1. Fig 3.7.4 shows what happened when a crystal of potassium manganate(VII) was placed in a beaker of water and left for 24 hours.

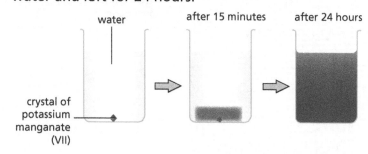

FIG 3.7.4

Explain why the appearance of the water changed.

> **Fun fact**
>
> Substances move into and out of living cells by diffusion.

Key terms

kinetic energy form of energy associated with movement

random motion unpredictable movement in all directions

concentration the number of particles of a substance per unit volume

concentration gradient the difference in the concentration of particles of a substance in one place compared to another place

53

Osmosis

We are learning how to:

- describe how water moves through a differentially permeable membrane during osmosis.

Osmosis 》》》

Osmosis is a special kind of diffusion that is concerned only with the movement of water molecules. Like other substances, water diffuses along a concentration gradient from a place where water is in high concentration to a place where it is in low concentration.

This can sometimes be a little confusing, because water is in a higher concentration in a dilute solution and in a lower concentration in a concentrated solution. What we are saying, therefore, is that under suitable conditions, water will move from a dilute solution, making it more concentrated, to a concentrated solution, making it more dilute.

A **differentially permeable membrane** is a membrane with holes that are large enough to allow small molecules like water to pass through, but small enough to prevent the movement of large molecules.

Fig 3.8.1 shows what happens if we separate pure water from a solution of a compound by a differentially permeable membrane. Notice that the water molecules move in both directions through the differentially permeable membrane. However, more water molecules pass from the dilute solution to the concentrated solution than pass in the other direction. Eventually the concentration of the solution reaches an **equilibrium** situation and remains unchanged.

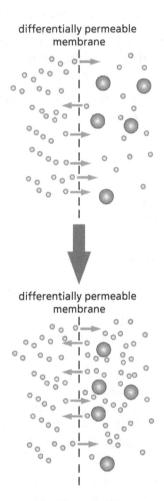

differentially permeable membrane

differentially permeable membrane

FIG 3.8.1 Movement of particles across a differentially permeable membrane

Activity 3.8.1

Investigating osmosis

Here is what you need:

- Fresh fruit – unripe, mature mango or paw-paw
- Knife
- Shallow dish
- Sugar.

Here is what you should do:

1. Peel the fruit with a knife and cut the papaya in half, remove the seeds. If you are using a mango, cut off one side of the mango.

2. Cut the outside of each half to make a flat surface for it to stand on.

FIG 3.8.2

3. Stand the two halves of the fruit on their flat parts in a shallow bowl of distilled water so that the well is at the top. In one of the wells, just cover the bottom with a layer of sugar but do not fill the well.

4. Do not put any sugar in the well of the other half of the fruit.

5. Leave the halves of fruit in the water for 30 minutes.

6. After 30 minutes, record your observations.

7. Explain your observations in terms of osmosis.

In Activity 3.8.1 the sugar dissolves in one half of the fruit, forming a sugar solution. The concentration of water in the dish is greater than the concentration of water in the sugar solution, so water passes from the dish into the fruit, through the fruit cells, by osmosis.

Water passes into and out of all living cells by osmosis. The process of osmosis controls the concentration of substances in the cell.

Key terms

osmosis diffusion that involves water molecules

differentially permeable membrane a membrane that allows some particles to pass through it but not others

equilibrium a situation in which the movement of particles in one direction is equal to the movement of particles in the opposite direction

Check your understanding

1. Visking tubing is a differentially permeable membrane. Fig 3.8.3 shows a bag made of Visking tubing, filled with 20% sugar solution and then suspended in a beaker of distilled water.

a) Predict what will happen to the level of liquid if the glass tube of the apparatus is left to stand for 30 minutes.

b) Explain your answer to **a)**.

c) Predict what will happen if the same apparatus is used, but distilled water is placed in the Visking bag and sugar solution in the beaker.

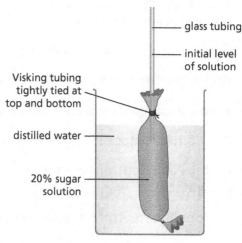

FIG 3.8.3

Review of More about matter

- An element cannot be broken down into any simpler substance.

- There are 94 naturally occurring elements.

- Atomic symbols are a shorthand way of representing elements.

- Each atomic symbol consists of one or two letters of the alphabet derived from their names.

- The atomic symbols of some elements are derived from their Latin names.

- The Periodic Table contains all of the known chemical elements arranged according to their structure.

- A row of the Periodic Table is called a period.

- A column of the Periodic Table is called a group.

- The symbols of metallic elements are to the left and in the centre of the Periodic Table.

- The symbols of non-metallic elements are to the right of the Periodic Table.

- Elements along the line where metals and non-metals meet often have a mixture of metallic and non-metallic properties and are called semi-metals.

- Metals are generally hard and have high melting points and boiling points, are good conductors of both heat and electricity and some can be polished to a high lustre.

- Non-metals are often soft and have low melting points and boiling points, are poor conductors of heat and, with the exception of a form of carbon called graphite, they are poor conductors of electricity.

- Groups of the Periodic Table include the alkali metals, the transition elements, the halides and the noble gases.

- An atom is the smallest part of an element that can take part in a chemical reaction.

- Atoms are the building blocks from which all substances are made.

- John Dalton proposed an atomic theory involving atoms 200 years ago.

- An atom is composed of sub-atomic particles – protons, neutrons and electrons.

- Protons and neutrons are found in the nucleus of an atom and electrons in shells around it.

- Protons and neutrons have a relative mass of 1 while the relative mass of an electron is 0.

- Protons and electrons carry equal but opposite charges.

- Atoms of the same element always have the same number of protons but may have different numbers of neutrons.

- Atoms of the same element which have different numbers of neutrons are called isotopes.

- The kinetic theory is concerned with the arrangement and movement of particles in a substance.

- Diffusion is the movement of particles from a region of higher concentration to a region of lower concentration along a concentration gradient.

- Osmosis is a special kind of diffusion that involves water molecules. During osmosis water molecules move from a more dilute solution to a more concentrated solution through a differentially permeable membrane.

Review questions on More about matter

1. a) Give the symbols of the following elements:

sodium   nitrogen   bromine   neon   beryllium

b) Which elements are represented by the following symbols?

He   K   F   Ca   A

2. Fig 3.RQ.1 shows the positions of the first 20 elements in the Periodic Table.

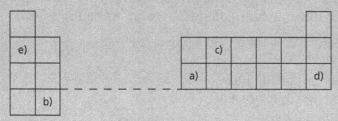

FIG 3.RQ.1

Which elements occupy positions a) to e)?

3. a) In which part of the Periodic Table are:

i) most metals?   **ii)** most non-metals?

b) In the Periodic Table what name is given to:

i) a column?   **ii)** a row?

c) i) What is a semi-metal?

ii) Name one example of a semi-metal.

4. An atom is composed of sub-atomic particles. Which particles:

a) are equal in mass?

b) carry equal but opposite charges?

c) are found in the nucleus of an atom?

5. A student placed one small crystal of the dye gentian violet on the surface of cold water in a beaker. Fig 3.RQ.2 shows what happened over the following hour.

gentian violet crystal

water

FIG 3.RQ.2

a) By what process do particles spread through water?

b) Describe what happened to the crystal during the hour.

c) How can you tell that the dye particles are distributed evenly throughout the solution?

d) Predict what would have happened if the same crystal had been placed in a large beaker containing ten times as much water.

e) Predict what would happen if the same crystal had been placed in a beaker of warm water.

---

6. A student carried out an investigation in which she placed 1 cm and 2 cm cubes of clear agar gel in potassium manganate(VII) solution for 30 minutes. After this time she removed the cubes and cut them in half. Fig 3.RQ.3 shows what she observed. The purple colour shows how far the potassium manganate(VII) solution had reached.

a) By what process has the potassium manganate(VII) moved into the blocks of agar gel?

b) Explain why the potassium manganate(VII) has reached the centre of the 1 cm block but not the centre of the 2 cm block.

1 cm block   2 cm block

FIG 3.RQ.3

---

7. Winnie added some berries to sugar solution to make jam. Here is what happened to the berries over the next couple of hours.

a) Explain what happened.

b) Predict what would happen if Winnie made her sugar solution twice as strong.

c) Suggest how the fruit could be restored to its original size.

FIG 3.RQ.4

8. The diagram represents an atom of lithium.

a) Name the sub-atomic particles X, Y and Z.

b) In terms of the sub-atomic particles present, explain why an atom is always neutral.

c) From the position of lithium in the Periodic Table, deduce whether it is a metal, a semi-metal or a non-metal.

d) Give one property of all metals which is absent from all non-metals.

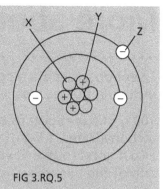

FIG 3.RQ.5

9. Part of the Periodic Table is shown here. Use only the elements shown in the diagram to answer the questions that follow.

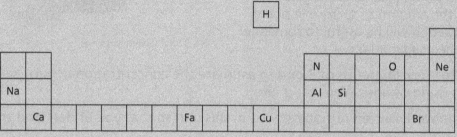

FIG 3.RQ.6

Give the symbol and the name of each of the elements described below.

a) A metallic element used to make pipes and wires.

b) A non-metallic element essential for combustion.

c) A metallic element which has a low density and is used to make cooking utensils.

d) A non-metallic element used in coloured strip lighting.

e) An element that has a mixture of metallic and non-metallic properties.

10. Gases diffuse at different rates. Fig 3.RQ.7 shows what happens when an inverted beaker containing hydrogen is placed over a porous pot containing air. The porous pot allows gases to diffuse through it. The porous pot is connected to a manometer which compares the gas pressure in the pot with atmospheric pressure.

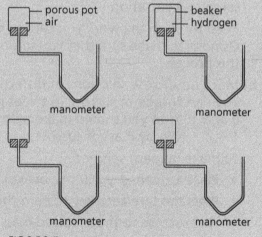

FIG 3.RQ.7

a) Shortly after the beaker of hydrogen is placed over the porous pot, the pressure in the porous pot becomes greater than atmospheric pressure. Suggest a reason for this.

b) When the beaker is removed the pressure inside the porous pot quickly becomes less than atmospheric pressure. Suggest a reason for this.

c) Predict what will happen to the pressure inside the porous pot if the apparatus is left for a short time. Explain your prediction.

More about matter

Sweet potatoes are a popular food in the Caribbean. They are eaten in lots of different ways including boiled, mashed, baked and as fries.

100 g of sweet potato contains about: 20 g of carbohydrates, 1.6 g of protein and 0.1 g of fat. It is also a good source of vitamins A, B-6, B-12, C and D, and of the minerals calcium, iron and magnesium.

As the name 'sweet potato' suggests, a proportion of the 20 g of carbohydrates is present as sugar (sucrose). You have been asked to use your knowledge of osmosis to estimate the amount of sucrose in sweet potatoes. This information will be useful to someone wanting to reduce their sugar intake.

FIG 3.SIP.1 Sweet potatoes

1. You are going to work in groups of 3 or 4 to estimate the concentration of sucrose in sweet potatoes using osmosis. Your tasks are:

   - To revise the work carried out on osmosis in this unit so that you understand the process.
   - To prepare sucrose solutions of different concentrations.
   - To measure the mass gain or mass loss when sweet potato chips are soaked in sucrose solutions of different concentrations.
   - To draw a graph of mass gain/loss against concentration and use it to deduce the sucrose concentration in sweet potatoes.

   a) Osmosis is a special kind of diffusion involving water molecules. When solutions of different concentrations are separated by a differentially permeable membrane there is a net movement of water molecules from the less concentrated solution to the more concentrated solution. Eventually the solutions will have the same concentration.

   Read through the work you carried out in Unit 3.6 Osmosis to make sure that you understand this process.

   b) Make up sucrose solutions of different concentrations. The percentage of sucrose in sweet potato is between 0 and 10% so you should make up solutions of 0%, 1%, ... 10% by mass of sucrose.

   FIG 3.SIP.2 Making up solutions

   To make up each solution:
   - Place a clean, dry 250 cm³ beaker on a balance
   - Press the tare/zero key to zero the display
   - Add sucrose to the required mass
   - Top up with distilled water until the display reads 100 g

   Place some cling wrap over the solution to prevent water evaporating or being absorbed from the air before you use it. Make sure you label each beaker with the concentration of sucrose solution it contains.

c) From one or more peeled sweet potatoes cut 55 chips of similar thickness.

The chips should not be too thick. This shape provides a large surface area for osmosis to take place.

d) Use a spreadsheet to record the following data:

- Weigh a batch of five chips and record their mass. Place them in 0% sucrose solution. Repeat this for each concentration up to 10% sucrose solution.

- Leave the chips overnight. The following day remove each batch of five chips from each solution, wipe them to remove surface solution and then reweigh them. Record their mass.

- Use the spreadsheet to calculate the change in mass and the percentage change in mass for each batch.

FIG 3.SIP.3 Sweet potato chips

e) Use the data you have collected to plot a graph of percentage mass gained or lost on the *y*-axis against concentration of sucrose in the solution on the *x*-axis.

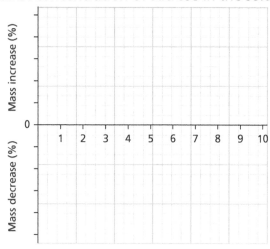

FIG 3.SIP.4 Grid for plotting data

Draw the straight line of best fit through your points and find the concentration at which there would be no increase or decrease in the mass of sweet potato chips. This will be your estimate of the concentration of sucrose in sweet potatoes.

f) Prepare a PowerPoint presentation in which you should focus on the results you obtained and make suggestions as to how people might use your results. You might pose questions like:

- What is the daily recommended intake of sugar?

- How many spoonfuls of sugar does a typical sweet potato contain?

- Should sweet potato be something you eat every day?

As part of your presentation you might hold up a sweet potato and an equivalent mass of sugar to emphasise how much it contains.

Creating an attractive version of the Periodic Table

The Periodic Table shown in lesson 2 of this unit provides information about over 100 known elements, but the unit itself is only concerned with the first 20 of these elements.

Your science teacher is concerned that the full version of the Periodic Table contains too much information for many students to comprehend. This is having a negative effect on student learning, which is affecting their appreciation of the whole unit content.

Your teacher wants you to use your understanding of the elements and their arrangement in the Periodic Table to produce a shorter version of the Periodic Table that focuses on the first 20 elements. It needs to be attractive to stimulate student interest.

1. You are going to work in groups of 3 or 4 to create a shorter version of the Periodic Table as a wallchart that can be used as a teaching aid. The tasks are:

    - To review the layout of the Periodic Table in the unit.
    - To identify the first 20 elements.
    - To consider what information is given about each element in the Periodic Table.
    - To decide how elements should be laid out in a shorter version and what information should be given about each element.
    - To consider if any additional information can usefully be added to the shorter version.
    - To produce a prototype of your Periodic Table for your teacher to comment on.
    - To amend your design based on feedback and to produce a wallchart version.
    - To produce an information sheet that can be used alongside your shorter Periodic Table.

    a) Look back through lesson 2 of this unit, and particularly at the full Periodic Table. Make sure you appreciate the significance of the periods and groups in terms of atomic structure.

    b) Identify the first 20 elements starting with hydrogen. How many groups and how many periods will your shorter version require?

       Remember that the layout of the Periodic Table is related to atomic structure, particularly the arrangement of electrons surrounding the nuclei of atoms. You must keep the existing groups and periods and cannot reorganise the arrangement to give a different shape.

    c) In the full version of the Periodic Table, for each element there is:

       - its atomic number
       - its atomic symbol
       - its name.

       Is all the information necessary for each element? Are atomic number, symbol, element and relative atomic mass covered in the unit?

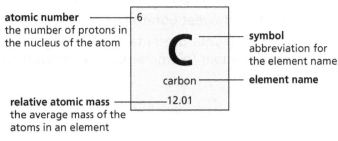

FIG 3.SIP.4 Information about each element

Is there any other information that can be usefully added? For example:

- lesson 3.3 is concerned with metals and non-metals
- some elements are solids, some liquids and some gases at room temperature.

You want your shorter Periodic Table to contain lots of useful information; however, if you add too much there is a danger it will seem too complicated and defeat the purpose for which it is being created.

d) Your next task is to produce a prototype, which will be commented on by your teacher. This should incorporate all the decisions you have made regarding structure and content.

Scientific illustrations like charts and graphs often make use of colour to improve clarity and increase user friendliness. Can you make use of colour in this way? Note that excessive use of colour can sometimes reduce the effectiveness of an illustration.

e) How was your shorter Periodic Table received by your teacher? How are you going to modify your design in the light of the comments you received?

f) To be an effective teaching aid, your shorter Periodic Table needs to be large enough to be seen by the whole class. Measure some of the wallcharts in your classroom and decide on an appropriate size. Your teacher may have some large sheets of paper that you can use or you may have to ask the Head of Art at your school for a sheet.

FIG 3.SIP.5 Wallcharts need to be big enough for the people at the back of the class to read

Lay out your Periodic Table to make the best use of the space available. There is no point in drawing an A4-sized chart on a sheet of A0 paper. Do some careful measuring to make sure the element boxes on your table are equal in size. Would the paper be best used portrait or landscape?

g) Once you have completed you simplified Periodic Table, your final task is to produce an information sheet to accompany it. Your information sheet should contain details such as:

- what information is given for each element
- the significance of colour in relation to metal/non-metal or state.

Unit 4: Human nutrition

We are learning how to:

- describe the digestive system
- understand the important features of human nutrition.

Human nutrition >>>

A diet is all of the things that a person eats and drinks.

Food may be placed into groups according to the nutrients it contains. These nutrients are:

- carbohydrates
- fats
- proteins
- vitamins
- minerals.

Two other important components of a diet are fibre and water.

FIG 4.1.1 Food enters the body through the mouth and passes into the digestive system

The process of digestion allows nutrients to be absorbed by the body. Once in the body, they are used to provide energy and to supply what is needed for the growth and repair of tissues.

FIG 4.1.2 There are many different foods to choose from

Analysing my diet

You do not need any equipment or materials for this activity.
Here is what you should do:

1. Copy Table 4.3.1.

Meal	'Go' foods (carbohydrates and fats)	'Grow' foods (proteins)	'Glow' foods (vitamins and minerals)
Breakfast			
Lunch			
Dinner			

TABLE 4.3.1

2. Think about what you had to eat yesterday. For each meal, write down in your table the five foods that you ate most of.

3. Alongside each food, place ticks to show whether it is mainly a 'go', a 'grow' or a 'glow' food. Each meal should get five ticks. If you think the food for a meal is all in one group, then all five ticks go to that group. If you think the food for a meal is a mixture of two or even three groups, then allocate the ticks accordingly. Add any snacks you had during the day to the table as well.

4. From which group did you eat most foods?

5. From which group did you eat fewest foods?

Fun fact

The energy content of food is expressed both in kilojoules (kJ) and in kilocalories (kcal). One calorie is equivalent to 4.18 joules.

The calorie is the old unit of energy. It is where terms like 'calorific value' and 'calorie counter' come from.

Key terms

diet the food you eat

balanced diet a diet that contains all the different nutrients that our body needs to stay healthy

nutritional information information about the nutrients contained in a food

Check your understanding

1. Look at Fig 4.3.2 again.

   a) Make a list of the nutrients present in this food.

   b) Is this food most likely to be cheese, meat or rice? Explain your answer.

Diet, activity and age

We are learning how to:

- recognise the importance of a balanced diet
- appreciate how diet relates to activity and age.

Energy needs »

To keep your body working, you need to obtain **energy** from your food. You need energy to breathe, to keep your heart beating and to drive the countless chemical processes that go on all the time throughout your body.

The rate at which these processes take place is called the **basal metabolic rate** (BMR). BMR depends on age and gender, so it varies from person to person. A typical value for an adult is 7000 kJ per day. The total amount of energy a person needs each day will be the sum of their BMR plus additional energy related to how active they are (**activity-related energy**):

total energy needed = BMR + activity-related energy

People who do manual work need a considerable amount of activity-related energy. They need to eat 'go' foods, which are rich in carbohydrates, to give them energy.

Working at a desk all day requires much less activity-related energy. Workers who sit down most of the day need to eat fewer 'go' foods than manual workers.

People who are still growing need 'grow' foods that contain proteins to build new tissues. A fully grown person only needs proteins to repair existing tissues. Teenagers need more 'grow' foods than adults.

FIG 4.4.1 Some people's jobs require a lot more energy than others

FIG 4.4.2 A person who is growing needs more 'grow' foods than a person who is fully grown

A pregnant woman needs a healthy, balanced diet to ensure that her unborn child has all the nutrients it needs to develop.

'Glow' foods such as fruit and vegetables keep us healthy. People sometimes do not eat enough 'glow' foods.

FIG 4.4.3 Everyone should eat several portions of fruit and vegetables each day no matter what job they do or how old they are

Activity 4.4.1

Advising people on their diet

This activity is a role play in which people are given advice about their diets. Working in your groups:

1. Choose one student who is going to be the dietician. This is a professional person who advises people about their diets.

2. The remaining students will pretend to be people of different ages with different occupations, for example a 20-year-old shop assistant or a 70-year-old retired person.

3. Each person should think of some difficult questions to ask the dietician.

4. The dietician should give each person advice about their diet according to their activity and age.

5. Swap roles so that everyone in the group gets a turn at being the dietician.

Check your understanding

1. Table 4.4.1 shows the amount of energy required each day by three people: A, B and C.

Person	Amount of energy required each day (kJ)
A	15 000
B	9 500
C	11 000

TABLE 4.4.1

a) Which of these people is likely to be:
   i) a teenage girl?
   ii) an office worker?
   iii) a labourer on a building site?

b) Estimate the amount of energy that person A needs each day as a result of their job.

> **Fun fact**
>
> The amount of energy different foods contain is measured by a food calorimeter. A known mass of the food is burned in oxygen and the amount of heat produced is measured.

Key terms

energy what the body needs to function, usually obtained from food

basal metabolic rate rate at which the basic functions of the body, such as breathing or the heart beating, take place

activity-related energy energy related to how active a person is

The digestive system (1)

We are learning how to:

- outline the basic structure of the digestive system and the functions of each part
- identify parts of the digestive system.

Getting digestion started 》》

The **digestive system** obtains nutrients and water from the foods that we eat.

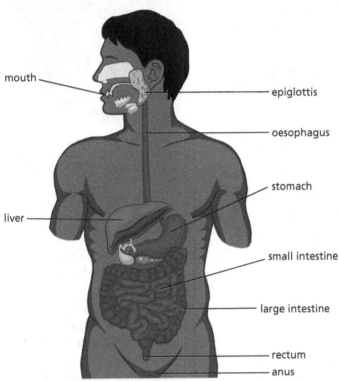

FIG 4.5.1 The digestive system

mouth
epiglottis
oesophagus
stomach
liver
small intestine
large intestine
rectum
anus

Digestion begins in the **mouth**. Food is chopped into smaller pieces by the pointed teeth and crushed by the flat teeth. Breaking the food into smaller pieces allows digestive enzymes to penetrate it more easily.

It is important to keep our teeth clean and healthy.

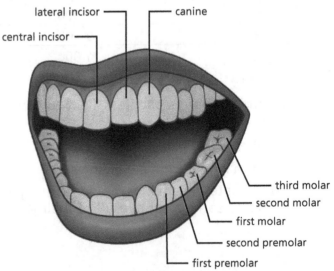

lateral incisor
canine
central incisor
third molar
second molar
first molar
second premolar
first premolar

FIG 4.5.2 There are four different kinds of teeth in our mouths

Activity 4.5.1

Investigating teeth

Here is what you need:

- Toothbrush
- Toothpaste
- Small mirror.

Here is what you should do:

1. Clean your teeth before starting this investigation.

2. Look carefully at the teeth right at the front of your mouth. What shape are they?

3. Make a drawing of your front teeth showing the shape.

4. By thinking about the shape, can you suggest what job this kind of tooth does?

5. Look along your front teeth to the side of the mouth until you see a tooth that is a different shape.

6. Make a drawing of your side teeth showing the shape.

7. By thinking about the shape, can you say what job this kind of tooth does?

8. Look at the teeth at the back of your mouth.

9. Make a drawing of your back teeth showing the shape.

10. By thinking about the shape, can you say what job this kind of tooth does?

The saliva released in the mouth contains enzymes known as carbohydrases. These break down the food by converting carbohydrates into simple sugars such as glucose.

Check your understanding

1. Look carefully at these two teeth.

a) 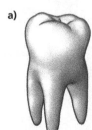   b)

FIG 4.5.3

State what type each tooth is and explain how the shape of each tooth is suited to its role in digestion.

> **Fun fact**
>
> People get two sets of teeth during their lives. When they are very young they have a set of 20 milk teeth. As they grow older, these are replaced by a set of 32 permanent teeth.

Key terms

digestive system the parts of the body that work together to process the food you eat

mouth where food enters the digestive system

The digestive system (2)

We are learning how to:

- outline the basic structure of the digestive system and the functions of each part
- identify parts of the digestive system.

Moving digestion along 》》

When you swallow, food from the mouth passes down the oesophagus into the **stomach**.

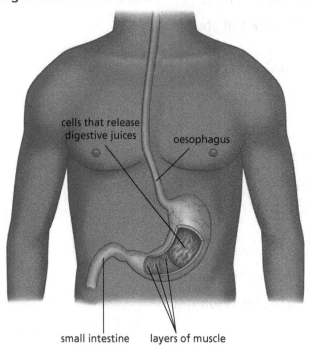

FIG 4.6.1 The stomach is a muscular bag that is about the size of your fist when empty but can expand to ten times this size when full of food

The wall of the stomach contains cells that release digestive juices. These include hydrochloric acid, which makes the stomach contents very acidic. Stomach juices also contain enzymes called proteases. Proteases act on proteins and break them down into chemicals called peptides.

The movement of the stomach muscles continues to break up the food and mixes it as digestive enzymes are added.

The amount of time that food spends in the stomach depends on what type of food it is. Typically, a carbohydrate meal like porridge passes through to the intestines in less than an hour, while a mixed meal containing proteins and fats may remain in the stomach for several hours before passing to the intestines.

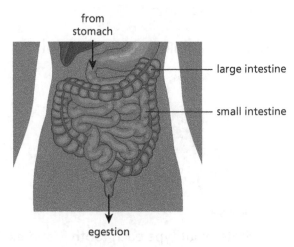

FIG 4.6.2 The intestines

The **small intestine** is about 7 metres long in an adult. The digesting food passes into the small intestine from the stomach and is moved along it by the rhythmic contraction of the muscular walls. This rhythmic contraction is called peristalsis.

In the first part of the small intestine – the duodenum – the cells produce enzymes that break down the food further.

Digestion continues in the second part of the small intestine – the ileum – and this is also where most of the absorption of nutrients takes place.

The **large intestine** is about 1.5 metres long in an adult. No further absorption of nutrients takes place in the large intestine. It has the important function of absorbing water back into the body.

The remains of the food in the large intestine contain mostly cellulose from plants. The body cannot digest these remains. They are egested from the body as faeces.

Activity 4.6.1

Looking at a model of the digestive system

Here is what you need:

- Model of the digestive system.

Here is what you should do:

1. Look carefully at the model. Identify the different parts that have been described.
2. Try to visualise where the various parts are in your body.

Check your understanding

1. In which part of the digestive system:

    a) is food chopped into small pieces and crushed?
    b) is water reabsorbed?
    c) does most absorption of nutrients take place?
    d) is acid added to the food?
    e) does the chemical breakdown of food start?

Fun fact

The epiglottis is a small flap of cartilage in the throat. When you swallow it moves down and blocks the entrance to the larynx so that food is directed down the oesophagus.

Key terms

stomach muscular bag-like organ, a part of the digestive system, where food is mixed with enzymes

small intestine the part of the digestive system where food is broken down further and nutrients are absorbed

large intestine the part of the digestive system where water is absorbed back into the body

The process of digestion

We are learning how to:

- explain how humans obtain nutrients from food
- describe the physical and chemical changes in food during digestion.

Breaking down food ⟩⟩⟩

During digestion food goes through both **physical changes** and **chemical changes**.

- The physical changes involve breaking the food down into small pieces through biting, chewing in the mouth and crushing in the stomach.

- The chemical changes involve the action of enzymes to break the large molecules in food into smaller molecules that can be absorbed easily by the body.

Both sets of processes are important. Biting and chewing the food does not release nutrients in food, but it does turn the food into a form in which chemical reactions can take place more efficiently.

Activity 4.7.1

Particle size and the rate of a chemical reaction

Your teacher will carry this out as a demonstration.

Here is what you need:

- Balance
- 250 cm³ beaker
- Measuring cylinder
- Stopwatch
- Limestone – one large lump of mass about 2 g
- Calcium carbonate – fine powder
- Hydrochloric acid of concentration 1 mol dm⁻³.

 SAFETY
Be careful when using acid. Follow local regulations.

Here is what you should do:

1. Weigh your lump of limestone (limestone is an impure form of calcium carbonate). It should be about 2 g.

2. Place the limestone in a 250 cm³ beaker.

3. Measure 100 cm³ of hydrochloric acid of concentration 1 mol dm⁻³ in the measuring cylinder.

4. Add the hydrochloric acid to the beaker and start the stopwatch at the same time.

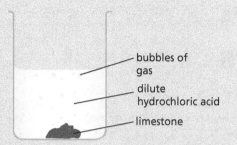

bubbles of gas

dilute hydrochloric acid

limestone

FIG 4.7.1 Limestone in hydrochloric acid

5. Continue observing and timing until the reaction is complete. This will be when all of the limestone has dissolved and no more bubbles are given off.

6. Weigh out the same mass of calcium carbonate powder as the mass of your piece of limestone.

7. Repeat steps 2 to 5.

8. Copy and complete Table 4.7.1 with your results.

Form of calcium carbonate	Appearance	Time taken for reaction to reach completion
single piece	large lump	
fine	powder	

TABLE 4.7.1

9. What evidence is there from your observations that breaking food into smaller pieces will increase the rate at which enzymes can break it down?

Not only does chewing food into small pieces make it easier to swallow but it also makes it easier for digestive enzymes to act on the food.

Check your understanding

1. Fig 4.7.2 shows two different sorts of sugar.

a)

b)
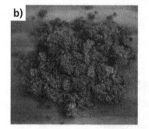

FIG 4.7.2

If equal masses of the sugar were added to cups of coffee and stirred, which sugar would dissolve first? Explain your answer.

Key terms

physical changes changes that take place as a result of physical processes such as chewing

chemical changes changes that take place as a result of chemical reactions such as action of enzymes on food

Food tests – starch and simple sugars

We are learning how to:

- explain how humans obtain nutrients from food
- test for different components in foods.

Testing for starch ⟩⟩⟩

Food can be tested for carbohydrates, either in the form of **starch** or as **simple sugars** such as glucose.

You might recall the test for starch from the work you carried out on photosynthesis.

When a few drops of **iodine solution** are added to a food containing starch, it turns a blue-black colour.

Testing for simple sugars

To test for the presence of a simple sugar, place a small sample of the food into a test tube containing **Benedict's reagent**. Heat the test tube for a few minutes in a hot water bath. If a simple sugar is present, the mixture will turn brick red.

FIG 4.8.1 Positive test for the presence of starch

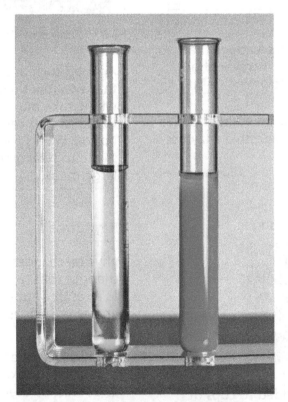

FIG 4.8.2 Positive test for the presence of simple sugars. The test tube on the left contains Benedict's reagent and the one on the right shows Benedict's reagent with a simple sugar after heating

Activity 4.8.1

Testing for starch and simple sugars

Here is what you need:

- Starch solution
- Benedict's reagent
- Iodine
- Water
- Test tubes
- Reagents for protein test (Biuret reagent)
- Food samples.

⚠️ **SAFETY**
Be careful when using chemicals. Follow local regulations.

Here is what you should do:

1. Your teacher will provide you with samples of some foods.

2. Test each food for the presence of starch and simple sugars.

3. Arrange your results in a table like Table 4.8.1. Place a tick or a cross in each column next to each food as appropriate.

Food sample	Starch present	Simple sugars present

TABLE 4.8.1

Check your understanding

1. You drop some iodine onto a food sample. The food sample turns blue-black. What does this tell you about the food?

2. How could you find out whether the food sample contains glucose?

A doctor who suspects that a patient might have diabetes can test their urine for the presence of glucose.

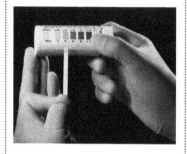

FIG 4.8.3 Testing urine for glucose

In the past they dipped their finger into the urine and tasted it to see if it was sweet! Nowadays, doctors use sticks coated in special chemicals.

Key terms

starch a form of carbohydrate found in foods

simple sugars a form of carbohydrate found in foods

iodine solution a solution containing iodine that is used to test for starch

Benedict's reagent a chemical used to test for the presence of simple sugars such as glucose

Food tests – proteins and fats

We are learning how to:

- explain how humans obtain nutrients from food
- test for different components in foods.

Food tests ▶▶

You can use food tests to find out which particular food components are present in different foods.

Testing for proteins

Biuret reagent is a mixture of sodium hydroxide solution and copper(II) sulfate solution. It is used to test for proteins.

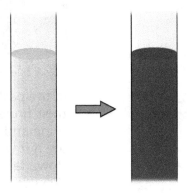

FIG 4.9.1 Positive test for the presence of proteins

When a sample of food is added to Biuret reagent, the reagent changes colour from blue to purple if protein is present.

Testing for fats

Fats are soluble in **ethanol** but insoluble in water. This provides a simple way to test for fats.

When a sample of food is shaken in a test tube with a small amount of ethanol, any fat it contains will dissolve in the ethanol.

If water is then added to the ethanol, any fat present comes out of solution as tiny globules that cause the ethanol to become cloudy.

FIG 4.9.2 Positive test for the presence of fats – note the cloudiness in the liquid at the top of the tube

Testing for proteins and fats

Here is what you need:

- Ethanol
- Water
- Test tubes
- Reagents for protein test (Biuret reagent)
- Food samples.

Here is what you should do:

1. Your teacher will provide you with samples of some foods.

2. Test each food for the presence of proteins and fats.

3. Arrange your results in a table like Table 4.9.1. Place a tick or a cross in each column next to each food as appropriate.

Food sample	Protein present	Fat present

TABLE 4.9.1

Check your understanding

1. A food gives positive tests with Biuret reagent and ethanol/water but negative tests with iodine solution and Benedict's solution.

   a) Which components are present in the food?

   b) Which components are absent from the food?

   c) Suggest what the food might be.

Key terms

Biuret reagent a mixture of sodium hydroxide solution and copper(II) sulfate solution used to test for the presence of proteins in food

ethanol a chemical in which fat dissolves

Review of Human nutrition

- Ingestion is the action of eating food.
- Digestion is what takes place as food passes along the digestive system.
- Absorption is the transfer of nutrients from the alimentary canal into the rest of the body.
- Assimilation is the use that the body makes of the absorbed nutrients.
- Egestion is the removal of undigested food from the body.
- 'Go' foods are rich in carbohydrates, which give the body energy.
- 'Grow' foods are rich in proteins, which the body needs to make new cells and repair damaged tissue.
- 'Glow' foods are rich in vitamins and minerals, which the body needs in small amounts.
- A balanced diet is one that contains sufficient, but not an excess of, all of the nutrients a person needs to remain healthy.
- Fibre or roughage has no nutrient value but it is an important component of the diet because it gives digesting food bulk, so it can be pushed along the alimentary canal.
- A person's diet should be related to their age and their level of activity.
- The basal metabolic rate, or BMR, is what a person needs to keep alive, even when they are at rest.
- The digestive system obtains nutrients and water from the foods that you eat.
- Digestion takes place in the alimentary canal. This consists of the mouth, oesophagus, stomach, small intestine and large intestine.
- Food is physically broken into pieces and crushed in the mouth by the act of chewing.
- Food is chemically broken down by chemicals called enzymes.
- Most absorption of nutrients takes place in the small intestine.
- Absorption of water takes place in the large intestine.
- Incisor and canine teeth are found at the front of the mouth. Their job is to cut and tear food into pieces.
- Premolar and molar teeth are found at the back of the mouth. Their job is to crush food.

Food tests

Food test for:	Reagent	Positive result
starch	iodine solution	blue-black colour
simple sugars	Benedict's reagent	blue to brick red
proteins	Biuret reagent	blue to purple
fats	ethanol followed by water	ethanol goes cloudy

Review questions on Human nutrition

1. a) Which type of teeth break food into smaller pieces?

   b) Why is the action of these teeth important for the process of digestion?

   c) What is the name of the group of chemicals that breaks food down during digestion?

   d) Name one example from this group of chemicals that acts on proteins.

   e) What nutrients are obtained from the breakdown of proteins?

2. Fig 4.RQ.1 represents the human digestive system.

   a) Which of the labelled parts is the:

   i) small intestine?

   ii) oesophagus?

   b) In which of the labelled parts:

   i) does digestion begin?

   ii) are most nutrients absorbed?

   iii) is acid added to food?

   iv) is most water absorbed?

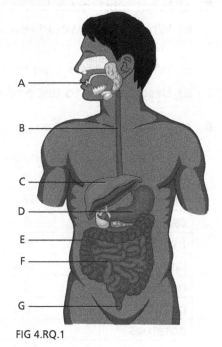

FIG 4.RQ.1

3. Fig 4.RQ.2 shows the nutritional information from a food package.

   a) Which of the three nutrient groups, carbohydrates, fats or proteins, does this food have:

   i) most of?

   ii) least of?

   b) What other important constituent of food, which has no nutritional value, does this food contain?

   c) By considering the information given, would you say that this food is a cereal, meat or cheese? Explain your answer.

**TYPICAL VALUES
NUTRITION INFORMATION**

	per 100 g	per 40 g	per 40 g***
Energy	1565 kJ/ 374 kcal	626 kJ/ 150 kcal	1222 kJ/ 291 kcal
Protein	11.0 g	4.4 g	8 g
Carbohydrate	60 g	24.0 g	38.4 g
(of which sugars)	1.1 g	0.4 g	14.8 g
Fat	8 g	3.2 g	8.6 g
(of which saturates)	1.5 g	0.6 g	3.9 g
Fibre	9 g	3.6 g	3.6 g
(of which beta glucan)	3.6 g	1.4 g	1.4 g
Sodium**	Trace	Trace	0.1 g
**Equivalent as salt	Trace	Trace	0.3 g

*** with 300 ml of semi skimmed milk

*Each serving (40 g) contains 47% of the 3 g of oat beta glucan suggested per day to help lower cholesterol as part of a varied and balanced diet and healthy lifestyle. Reducing cholesterol helps maintain a healthy heart.

FIG 4.RQ.2

4. a) Give three examples of 'go' foods.

b) How is the job a person does linked to the amount of 'go' foods he or she should eat each day?

c) What group of nutrients do 'go' foods contain?

d) How do 'go' foods provide the body with energy?

e) If a person eats more 'go' foods than their body needs, what happens to the extra nutrients?

5. a) What substance is iodine solution used to test for?

b) What change would you see when Benedict's reagent is heated with glucose in a water bath?

c) What reagent is used to test for the presence of proteins?

d) Describe how to test a food to find out if it contains fats.

6. Table 4.RQ.1 gives some information about 100 g samples of some different foods.

Food	Carbohydrates %	Fats %	Protein %
beef	0	28.2	14.8
bread	52	1.8	9
carrots	5.4	0	0.7
fish	0	0.5	16
jam	69.2	0	0.5
oranges	8.5	0	0.8
peanuts	8.7	49	28.1
potatoes	18	0	2
rice	86.7	1	6

TABLE 4.RQ.1

a) Name two other types of nutrient that are not shown in Table 4.RQ.1.

b) Which of the foods in Table 4.RQ.1 is the richest source of:

i) carbohydrates?

ii) proteins?

iii) fats?

c) Which two of the foods in Table 4.RQ.1 should be avoided by someone on a low carbohydrate diet?

d) Vegetarians do not eat meat or fish. Which two of the foods in Table 4.RQ.1 provide the best alternative sources of protein?

7. a) Explain why a professional sportsperson might need to eat twice as much 'go' foods as an ordinary person each day.

b) Predict what would happen to the body mass of a professional sportsperson if they continue to eat the same amount after they retire from the sport.

c) Explain your answer to part (b).

8. The following diagrams show the skulls and teeth of two animals, A and B.

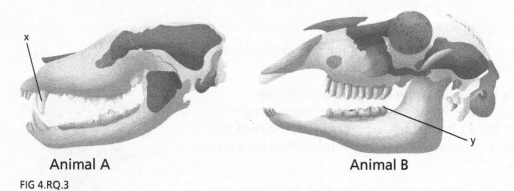

Animal A   Animal B

FIG 4.RQ.3

a) What type of tooth is:

   i) x?   **ii)** y?

b) State three differences between the type and arrangement of teeth in the skulls of the two animals.

c) Which animal is more likely to be a sheep? Explain your answer.

d) Which animal is more likely to be a wolf? Explain your answer.

9. Obese people who are unable to lose weight by following a weight-reducing diet can have a surgical procedure in which a gastric band is placed around the stomach.

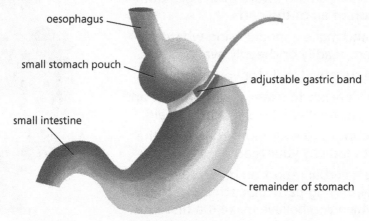

oesophagus

small stomach pouch

adjustable gastric band

small intestine

remainder of stomach

FIG 4.RQ.4

a) Suggest how a gastric band will help a person to lose weight.

b) Critics of the use of gastric bands claim that they deal with the result of overeating but not the cause. Explain whether you think this is a fair claim.

Human nutrition

Teachers often use diagrams and models in science to help student understanding.

Sometimes models can be bought but they are expensive. Bought models also sometimes omit parts a teacher thinks important, or include parts which are not necessary and which the teacher believes distract students.

Models made from commonly available materials are often just as good and much cheaper to make. Also they can have exactly as much detail as required.

Your teacher has asked you to use your knowledge of the alimentary canal to build a model which can be used as a teaching aid with students in the years to come.

1. You are going to work in groups of 3 or 4 to design and build a model of the alimentary canal. The tasks are:

   - To review the structure of the alimentary canal and make sure you are aware of the position and function of all of the parts.

   - To design and make a model using whatever materials are readily or cheaply available.

   - To prepare suitable labels for your model.

   - To ask your teacher to examine your model and assess how useful it will be as a teaching aid.

   - To make changes to your model based on the assessment made by your teacher.

   - To compile a verbal report on how you built your model, paying particular attention to those features which you believe make it a first-rate teaching aid. You should illustrate your report by taking pictures at different stages during the manufacture of your model. Your model should be exhibited when you make your report.

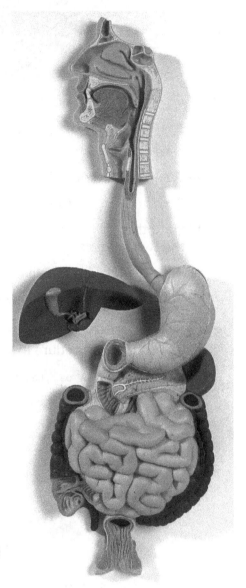

FIG 4.SIP.1 Bought model of the alimentary canal

   a) Look back through the unit and make sure that you are familiar with the different parts of the alimentary canal and the function of each.

   b) What parts of the alimentary canal will your model show? Start by making a list. The main ones are: mouth, oesophagus, stomach, small intestine, large intestine.

   You might want to include additional parts like the liver and the pancreas. You might want to add a little more detail, such as the different parts of the large intestine, i.e. colon, rectum, anus. Now is the time to decide on the features to be shown.

What materials are you going to need in order to build your model? What materials are available for you to use? Here is a model made by a student which might give you some ideas.

Think carefully about where you might source each of the parts you will need. Here are some thoughts to guide you.

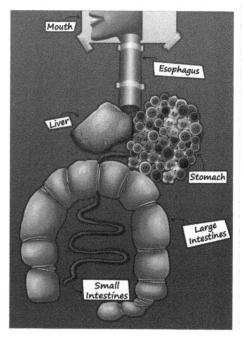

FIG 4.SIP.2 Model of the alimentary canal

- A mouth with teeth might be difficult to make. Is it possible to buy something like chattering teeth from a joke shop?

- The oesophagus is a long straight tube. Where can you find a suitable tube? The cardboard tube of a toilet role would be too short; the cardboard tube of a kitchen roll maybe?

- The stomach is a muscular sack. Maybe a small thick polythene bag would do? If you use a clear bag you could show some pretend food digesting which might add realism to your model?

- The small intestine needs a thin tube and the large intestine a thick tube. They need to be flexible.

You also need to take into account the scale of the different parts. They should be similar in relative size to the real thing, perhaps with the exception of the small intestine, which might need to be shortened to fit.

Are the parts of the model to be fixed together or will you build it in such a way that each part can be removed and shown to students?

Don't forget to take some photographs of your model at different times during the building. These will be useful when you make your report.

c) Your model will need some clear labels of a suitable size to be read from a distance of a couple of metres. Labels are often in the form of black writing on white card but you might want to experiment with different combinations of colours to improve clarity and make the model more attractive.

d) Once your labelled model is made, ask your teacher to cast a critical eye over it. He or she will be using this as a teaching aid so they are in the best position to determine whether the model will work well or not. Listen carefully to the comments and make some notes that you can use to guide any modifications you will make.

e) When your model has been modified to your satisfaction you need to prepare an oral report which you will deliver to the rest of the class. Describe the materials used and why you chose them. Discuss how you brought the different parts together using photographs to illustrate your talk.

Don't be afraid to discuss things that went wrong or did not quite turn out the way you expected them to. Negative results in science can often be useful in steering people in the right direction.

Designing a meal for people with special dietary requirements

Most people can eat whichever foods they choose but there are some people who cannot eat certain foods without experiencing severe discomfort and nausea.

People who suffer from coeliac disease are allergic to gluten and cannot eat any foods made from wheat or barley. This includes things like ordinary bread, cakes and biscuits.

People who are lactose intolerant cannot eat any foods made from cows' milk. These include milk, cream, yoghurt, butter and cheese.

FIG 4.SIP.3 Foods for people who have special dietary requirements

Your Principal has invited the school governors to lunch, but on acceptance of her invitation, she has discovered that one governor is coeliac and another is lactose intolerant. Your Principal has learned that you have recently been studying diet, so she has turned to you for help.

She wants you to design a three-course meal which can be eaten by all the governors. She also wants you to produce an attractive menu card that can be sent out to the governors in advance so they know what food will be served. This will reassure them that their special dietary requirements have been taken into account.

1. You are going to work in groups of 3 or 4 to devise a meal that takes into account special dietary requirements. The tasks are:

- To review the different food groups and how they combine to provide a balanced diet.
- To research into what gluten free foods and what lactose-free foods are available.
- To devise a three-course meal composed of foods from the different food groups in appropriate proportions.
- To discuss your meal plan with the home economics teacher at your school with a view to revising it based on their comments.
- To design and make a menu card providing a description of the dishes which can be sent out to participants prior to the meal.

a) Look back through the first half of the unit which is concerned with food groups and diet. Make sure you understand what is mean by a balanced diet and how this is achieved by combining different foods in suitable proportions.

b) Carry out some research into coeliac disease and lactose intolerance. Have a look in the 'Free from' section at your local supermarket. Make sure you understand these conditions and can identify the problem foods associated with each one.

c) You have been asked to serve a meal consisting of a starter, a main course and a dessert. How are you going to decide what to serve?

One way of exchanging ideas would be for each member of the group to write down their ideas for each course. You could then discuss ideas and come to a provisional decision about each course separately.

d) Do any of the meals you propose require foods that contain gluten or foods that contain dairy products?

Are there any suitable alternative foods that can be used? For example:

- Bread, cakes and biscuits made from gluten-free flour are available as is gluten-free pasta.

- Alternatives to dairy products like soya milk, soya ice cream, goat's cheese and dairy free spreads are available.

Revise your provisional plans by substituting gluten-free and dairy-free ingredients where necessary.

e) Ask the home economics teacher at your school to comment on your proposed meal. Does it provide a balance of different food groups and also avoid those foods that cannot be eaten by people with coeliac disease and lactose intolerance?

Was your teacher able to make any suggestions for improving your meal? If so, include these suggestions before you design your menu card.

f) Your final task is to create a menu card giving details of the proposed meal. Some examples are shown here.

FIG 4.SIP.4 Some examples of menus

Ignore the contents of the menus and focus on the different designs. Do they give you any ideas for your menu?

Like your meal, your menu card should be attractive so that people will really be looking forward to the meal before they arrive.

Your menu card should not have too much detail, but does need to indicate that you have considered the special dietary requirements of the guests. For example, if you plan to start the meal with vegetable soup, you could describe it as: 'Vegetable soup with gluten-free bread rolls'. Similarly, if you plan to end the meal with fresh fruit and ice cream, you could describe it as: 'Fresh fruit and dairy-free ice cream.'

Unit 5: Physical and chemical changes

We are learning how to:

- distinguish between physical and chemical properties
- understand the nature of physical changes
- understand the nature of chemical changes.

Physical and chemical changes »

The changes that happen to substances when they are treated in certain ways can be classified as physical changes or chemical changes.

Physical changes

Physical changes do not alter the nature of a substance but they may alter its appearance. Some physical changes are easy to reverse.

Heating an iron bar causes it to change colour. It becomes softer so it can be bent more easily. When the bar cools down however it is still made of iron. No permanent change has taken place.

When a current passes through a coil of wire, the wire becomes an electromagnet. However, when the current is turned off the coil is unchanged and is no longer a magnet.

Chemical changes

When a chemical change takes place, the nature of the substance is permanently changed. Chemical changes are generally difficult or impossible to reverse.

Rusting and burning are chemical reactions involving oxygen from the air. These are both chemical changes which produce new substances. For example, the properties of rust are very different to those of iron.

There are lots of examples of chemical changes in nature. Fruit ripening and the decay of fallen leaves are examples you might have seen.

FIG 5.1.1 Heating and cooling are physical changes

FIG 5.1.2 Magnetising is a physical change

FIG 5.1.3 Rusting and combustion are chemical changes

FIG 5.1.4 Fruit ripening is a chemical change

Pure and impure

In science we use the word 'pure' in a slightly different way to everyday language. A pure substance is a single substance, i.e. it contains no other substances or impurities. Distilled water is pure water.

FIG 5.1.5 Distilled water is pure water

In science we sometimes have to purify substances by removing impurities. We do this by using different separation techniques, depending on the nature of the impurities.

Check your understanding

1. State whether each of the following is a physical change or a chemical change.
   a) Diluting orange squash
   b) Frying an egg
   c) Melting chocolate
   d) Paint drying
   e) Burning charcoal
   f) Baking bread

Fun fact

Most of the substances we eat every day are not pure in the scientific sense. They are mixtures of substances but they are described as pure because they don't contain substances which are harmful to us.

Physical properties of matter

We are learning how to:

- identify some observable physical properties of matter
- categorise properties into quantitative and qualitative.

Physical properties ▶▶▶

All matter has **physical properties** that are observable.

Activity 5.2.1

Observing properties

Here is what you need:

- Materials such as sulfur, glass rod, stone, chalk, wire, candle
- Rubber bands
- Bunsen burner
- Magnet
- Paper
- Batteries
- Nail
- Coin
- Tongs
- Connecting wire
- Beaker.

 SAFETY
Wear eye protection for heating. Use correct procedure to light burner. Do not touch hot apparatus and materials with bare hands. Handle the thermometer carefully and do not let it roll off the desk.

Here is what you should do:

Record your observations in a table like this.

Material	sulfur	glass	stone	chalk, etc.
Colour				
Odour				
Does it conduct electricity?				
How hard is it?				
How elastic is it?				
Is the material attracted to a magnet?				
Does the material dissolve in water?				

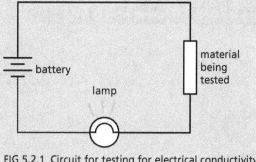

1. Build a circuit as shown on the right. Test for conductivity by putting each material into the circuit between the lamp and the battery to see whether the lamp lights up.

2. Test for hardness by scratching each material with a nail. The harder the material, the lighter the nail mark will be.

3. Stretch each material to test for elasticity.

FIG 5.2.1 Circuit for testing for electrical conductivity

The physical properties of a substance make it unique. These properties may be qualitative or quantitative.

Qualitative physical properties

Qualitative properties can be observed by using our five human senses, without changing the nature of the substance, for example:

- colour

- odour – the distinctive smell or aroma of the substance

- state – whether the substance is a gas, liquid or solid.

Quantitative physical properties

Quantitative properties give numerical information with the use of measuring instruments, for example:

- electrical conductivity – the ability to allow electricity to flow through

- heat conductivity – the ability to transfer thermal energy

- solubility – the ability of a substance to dissolve in another

- melting point – the temperature at which the change from solid to liquid occurs

- boiling point – the temperature at which the change from liquid to gas occurs

- hardness – the ability to withstand scratches

- elasticity – the extent to which a material returns to its original shape after stretching

- magnetism – whether a material is attracted to a magnet.

Fun fact

The strength of 1 cm³ of human bone is five times that of concrete.

Key terms

physical properties related to the appearance of a material

qualitative properties can be described by words only

quantitative properties can be measured

Check your understanding

1. Which physical properties cannot be described fully using just your senses? How would you go about describing them more completely?

2. Give three examples of quantitative properties and three examples of qualitative properties.

Chemical changes

We are learning how to:

- identify some features of a chemical change.

Effervescence >>>

Activity 5.3.1

Observing a simple reaction (1)

Your teacher will drop some food colouring into solutions of an alkali, an acid and hydrogen peroxide, then cover and shake each.

1. Discuss your observations.

Activity 5.3.2

Observing a simple reaction (2)

You should work in groups of 5 or 6.

Here is what you need:

- A bottle
- Dilute hydrochloric acid
- Baking soda
- A balloon
- A spatula
- An elastic band
- A funnel.

Here is what you should do:

1. Using the funnel, place two spatulas of baking soda into the balloon.

2. Half fill the bottle with dilute hydrochloric acid. Do not spill any.

3. Without allowing the baking soda to fall into the acid, stretch the opening of the balloon to fit the mouth of the bottle and seal with the elastic band.

4. Place the bottle on the table and stand away.

5. Lift the flopping end of the balloon so that the baking soda is poured into the bottle. What do you observe?

6. What type of material is baking soda? What happened when it combined with an acid? What could happen if you shook the combination?

FIG 5.3.1 A chemical change causes the balloon to inflate

Hydrochloric acid is an **acid** and baking soda is a **base**. The two react together and fizzing occurs as a gas is produced. The gas caused the balloon to inflate in Activity 5.3.2. The production of gases, or **effervescence**, indicates a chemical reaction. Shaking might cause the balloon to burst.

Activity 5.3.3

Observing a simple reaction (3)

You should work in small groups.

Here is what you need:

- Yeast
- Hydrogen peroxide
- A stirring rod
- A thermometer
- A test tube
- Gloves.

⚠️ **SAFETY**
Hydrogen peroxide is an irritant and a strong bleaching agent. Handle carefully. Reaction may be hot. Use gloves.

Here is what you should do:

1. Pour the hydrogen peroxide into the test tube, place the thermometer in the test tube and record its temperature.

2. Pour in the yeast and stir. What do you observe as the two substances combine?

3. What do you observe about the thermometer?

4. How does the outside of the test tube feel?

When hydrogen peroxide reacts with yeast, both fizzing and heat are produced. Heat also indicates a **chemical change**.

Colour change

Alkali, acid and hydrogen peroxide react differently with food colouring. The colour change indicates a chemical change.

Check your understanding

1. What is observed when an acid reacts with a base?

2. What happens when yeast and hydrogen peroxide combine?

3. Bleach, hydrogen peroxide and dilute hydrochloric acid are different materials. Which do you think is the least harmful?

4. Chemical properties describe how a substance behaves when combined with another substance. Do you think chemical properties can be observed via our senses?

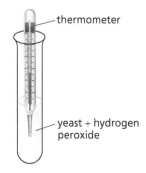

FIG 5.3.2 Measuring the temperature of yeast and hydrogen peroxide

Fun fact

Chemicals can be used to give some foods distinctive flavours.

Key terms

acid a substance that will react with a base

base a substance that will react with an acid

effervescence the fizzing that occurs when an acid substance reacts with a base substance to form a gas

chemical change a change that occurs when substances react together to form another substance

Comparing physical and chemical changes (1)

We are learning how to:

- classify changes as physical or chemical
- identify some differences between a physical and a chemical change.

The effect of heat on substances ≫

Different substances behave differently when heated.

Activity 5.4.1

Heating substances

Here is what you need:

- Salt
- Soil
- Lipstick
- Candle
- Brown or white sugar
- Tin lids
- Tongs
- Matches
- Tripod
- Gauze
- Bunsen burner
- Eye protection.

 SAFETY

Wear eye protection for heating. Use correct procedure to light Bunsen burner. Do not touch hot apparatus and materials with bare hands.

Here is what you should do:

1. Place each material individually on a tin lid and heat over a *gentle* flame for no longer than 2 minutes. Note your observations.

2. Allow each substance to cool, then observe again.

3. Record your observations in a table.

4. Discuss your observations in your group.

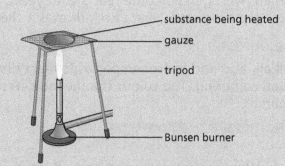

FIG 5.4.1 Apparatus for heating substances

substance being heated
gauze
tripod
Bunsen burner

You will have found that heating had little effect on salt and soil; the lipstick and the candle melted; and the sugar burned. The lipstick and the candle returned to the original substance on cooling. The sugar first melted and then it burned; it did not return to its original substance.

When a substance undergoes a change but can then return to the same substance with the same physical properties, it has undergone a physical change. This type of change is a **reversible change**.

When a substance undergoes a change and *cannot* return to the same substance with the same physical properties, it has undergone a chemical change. In all chemical changes, new products are formed. This type of change is usually an **irreversible change**.

Melting is a physical change. Burning is a chemical change. The fact that sugar burns is a chemical property of sugar.

Fun fact

While carbon is a black powder and hydrogen and oxygen are colourless gases, we all enjoy their chemical combination as sugar.

Activity 5.4.2

Writing a laboratory report for heating substances

Complete your laboratory report with the help of the following questions.

1. What was the aim of your experiment?

2. What apparatus and materials did you use?

3. How did you carry out the process of heating the materials? Draw a labelled diagram.

4. What safety precautions did you take?

5. How did you ensure your results were as accurate as possible? (For example, placing each substance on a clean lid and giving each substance the same heat intensity for the same time.)

6. What observations did you make? Did all the materials behave in the same way?

7. What conclusions can you draw from the experiment?

Check your understanding

1. From your experiment, two statements regarding the physical and chemical changes observed could be made. Copy and complete the following table. Allow space to add further rows to your table later.

	Physical change	Chemical change
1	The change ... be easily reversed.	The change ... be easily reversed.
2	No new substance is formed.	A new substance is formed.

2. Melting is a physical change: it can be reversed. From your knowledge of changes of state, name some other reversible changes.

Key terms

reversible change change where substance can go back to its original form

irreversible change change where substance cannot go back to its original form

Comparing physical and chemical changes (2)

We are learning how to:

- identify more differences between physical and chemical changes.

Differences between physical and chemical changes 》》

In the last experiment, you identified some differences between physical and chemical changes. In this experiment, you will identify further differences.

Activity 5.5.1

Comparing physical and chemical changes

Watch as your teacher demonstrates using aluminium and copper oxide.

Here is what you should do:

1. Observe each material. Describe its appearance.

2. Can you still identify each material when your teacher mixes them together? Would you be able to separate them?

3. Your teacher will then heat a small metal bowl of this mixture in a fume cupboard. Observe for a few minutes.

4. Can you explain what you observe?

5. Afterwards, look at the product and describe it. Could you recover the original substances?

FIG 5.5.1 **a)** Aluminium filings **b)** Black copper oxide powder **c)** The results of the reaction between aluminium and copper oxide

Aluminium is a silvery metal and copper oxide is a black powder. When the two substances were mixed, it was possible to identify and separate each one because of their colour and particle size.

When the two substances were heated, the appearance of the bright firework explosions indicated that a chemical change had occurred. The individual properties of the substances were lost. The heat caused a **chemical reaction** between the aluminium and the copper oxide. New substances – aluminium oxide and copper metal – with new properties were formed. The reaction is represented by the equation:

aluminium + copper oxide → aluminium oxide + copper

$$2Al \quad + \quad 3CuO \quad \rightarrow \quad Al_2O_3 \quad + \quad 3Cu$$

Check your understanding

1. You can now make further statements regarding physical and chemical changes. Add rows to the table you started in the last lesson, as below, and complete it.

	Physical change	Chemical change
1	The change can be easily reversed.	The change cannot be easily reversed.
2	No new substance is formed.	A new substance is formed.
		The new substance formed has … properties.
3	When substances are mixed the components … be easily identified.	When substances are combined the components … be easily identified.
4	After substances are combined the original components … be easily separated.	After substances are combined the original components … exist.

Fun fact

Gallium is an unusual metal that does not occur as a pure element in nature. A very interesting phenomenon about gallium is that it can melt in the palm of your hand.

Key term

chemical reaction
a reaction that takes place between two substances so that they combine to form a new substance or substances

Mixtures (1)

We are learning how to:

- describe some of the properties of mixtures
- identify some common mixtures.

Mixtures ▶▶

A **mixture** consists of two or more substances that are not chemically combined. Making a mixture is a **physical process** in which two components are shaken or stirred together. No chemical reaction takes place.

Mixtures may contain solids, liquids, gases or a mixture of these. Although there are many different mixtures, they all have certain general properties. You can use these properties to help you to recognise a mixture.

Water

Distilled water is pure water. It contains only water molecules but we seldom use distilled water outside the laboratory. More often the water we use in our everyday lives is a mixture of substances.

Around 70% of the Earth is covered in water and most of this is **sea water**.

Sea water tastes salty because there are many different substances, including sodium chloride (the same as table salt), dissolved in it. Sea water typically contains 3.5% dissolved solids. This means that each 1 kg of sea water contains 35 g of dissolved solids, but the actual amount varies from place to place.

- In places where large amounts of fresh water flow into the oceans, such as the Baltic Sea, the amount of dissolved solids is much lower.
- Conversely, where rivers flow into a land-locked sea, such as the Dead Sea, water evaporates and any dissolved solids remain and accumulate so the amount of dissolved solids is much higher.

Fresh water is also a mixture although it contains fewer dissolved solids than sea water.

FIG 5.6.1 The sea is a mixture of solids dissolved in water

FIG 5.6.2 Bottled mineral water also contains dissolved solids. Look at the label to see what minerals are dissolved in it

Fun fact

Blood is a mixture containing many different substances. The composition of the blood changes continually as it circulates around the body. For example, the blood leaving the lungs has a much higher proportion of dissolved oxygen than the blood entering the lungs.

Activity 5.6.1

Comparing the dissolved solids in some samples of water

Here is what you will need:

- Microscope slides × 4
- Dropper pipette
- Distilled water, tap water, mineral water, sea water.

1. Wash the microscope slides and dry them carefully using tissues to ensure they are clean.

2. Label the slides A, B, C, D and place them on a small board.

3. Near one end of slide A, place 4 drops of distilled water.

4. Repeat step 3 placing samples of tap water, mineral water and sea water on slides B, C and D, respectively.

5. Carefully place the board on a sunny windowsill and leave it overnight to allow the water from the samples to evaporate. Examine the slides the next day.

6. Draw a table listing the four slides and alongside each, comment on the amount of residue on each microscope slide.

FIG 5.6.3 When wood burns it reacts with the oxygen in air

Air

The atmosphere that surrounds the Earth consists of a mixture of gases that we call **air**.

Air does not have a constant composition. For example:

- Air is never dry, even in the desert. The amount of water vapour in the air varies from place to place.

- The proportion of carbon dioxide may be lower in the air above a forest, where trees are absorbing the gas for photosynthesis, and higher above a city where fuels are burned and extra carbon dioxide is formed.

- In some places, additional gases may be present in air in small amounts as pollutants. Sulfur dioxide and nitrogen oxides are pollutant gases.

The gases in air are not chemically combined together and therefore each gas will undergo the same chemical reactions when in air as it does when pure.

Check your understanding

1. a) What information given about sea water indicates that it is a mixture?

   b) What property of mixtures is demonstrated by burning things in air?

   c) Explain why distilled water is a compound while sea water is a mixture.

Key terms

mixture two or more substances which are physically combined but have not undergone a chemical reaction

physical process a process where two or more substances are physically mixed, but no chemical reaction has taken place

distilled water water containing no dissolved solids

sea water water containing dissolved solids

fresh water water used for drinking, cooking, etc. which contains fewer dissolved solids than sea water but is not chemically pure

air mixture of gases including oxygen, nitrogen and carbon dioxide

Mixtures (2)

We are learning how to:
- describe some of the properties of mixtures
- identify some common mixtures.

Crude oil >>>

Crude oil is a dark viscous liquid which is obtained from oil wells in different parts of the world. It is a mixture of chemicals called **hydrocarbons**. It was formed over millions of years by the decomposition of plant and animal remains deep under the ground.

The crude oils obtained from different places in the world may be similar in appearance but can be very different in composition, depending on the proportions of different hydrocarbons it contains.

For example, Arabian Heavy crude oil contains a much higher proportion of 'heavy' components, like fuel oil, than Arabian Light crude oil, which has more of the 'light' components such as naphtha (which is used to make gasoline) and kerosene (which is used to make aviation fuel).

FIG 5.7.1 Crude oil

FIG 5.7.2 Jamaica's $20 coin is composed of two alloys

Alloys >>>

An **alloy** is a mixture of a metal with other metals or non-metals. Many of the metals we use in everyday life are alloys.

For example, Jamaica's $20 coin was made using two alloys: cupronickel (an alloy of copper and nickel) in the centre and brass (an alloy of copper and zinc) around the rim.

Pure iron is relatively weak. Most of the items that we describe as being made of iron are really made of mild **steel**, which is much stronger. Steel is an alloy of iron with a very small amount of carbon.

Alloys provide a way of modifying the properties of a metal so that it is better suited to particular applications.

Aluminium has a low density but it is too weak to use for construction. Aircraft bodies are made of **duralumin**, which is an alloy of aluminium and copper. Duralumin has a low density like aluminium but it is much stronger.

FIG 5.7.3 Steel is an alloy of iron

FIG 5.7.4 Duralumin is an alloy of aluminium

Activity 5.7.1

Examining brass, copper and zinc

Here is what you will need:

- Small blocks of brass, copper and zinc
- Ruler
- Balance
- Electric circuit including cell, lamp and wires.

1. Examine the blocks of brass, copper and zinc. In what ways is brass similar in appearance and in what ways is it different in appearance to copper and to zinc?

2. How does the density of brass compare with that of copper and of zinc?

3. Include each of the blocks of metals, in turn, in a simple circuit containing a cell and a lamp. Is there any noticeable difference in the brightness of the lamp?

4. Comment on how well brass conducts electricity compared with copper and with zinc.

The characteristics of mixtures are summarised in Table 5.7.1.

FIG 5.7.5 Bronze axe heads

Characteristics	Mixture
Composition	Mixtures do not have an exact composition. The proportions of the different components can vary.
Separation	The components of a mixture can be separated using physical processes.
Properties	Each of the components in a mixture retains the same properties it has as a pure substance.
Appearance	The appearance of a mixture has features of all of the components from which it is formed.

TABLE 5.7.1

Check your understanding

1. Table 5.7.2 shows the composition of crude oil from some different locations.

Source	Naphtha %	Kerosene oil %	Fuel oil %
Algeria	30	40	30
Middle East	20	33	47
Nigeria	25	37	38
North Sea	23	35	42
Venezuela	1	19	80

TABLE 5.7.2 (*Source: Exxon-Mobile*)

a) Which sample of crude oil is richest in:

i) naphtha?   ii) kerosene oil?   iii) fuel oil?

b) Represent the data in Table 5.7.2 as a bar chart.

Key terms

crude oil mixture of hydrocarbons from which fuels are obtained

hydrocarbon chemical composed of hydrogen and carbon only

alloy mixture of a metal with other metals and non-metals

Compounds

We are learning how to:

- describe the properties of compounds
- identify some common compounds.

Compounds ⟩⟩

Chemical compounds are formed as a result of chemical reactions. During a chemical reaction **reactants** are used up and **products** are formed.

In order for two substances to undergo a chemical reaction it is not sufficient to simply mix them together.

For example, if magnesium metal is placed in the gas oxygen, nothing will happen. However, if the magnesium is heated then a chemical reaction takes place. Heat is often used to start chemical reactions.

FIG 5.8.1 Magnesium burns in oxygen to form magnesium oxide

magnesium + oxygen → magnesium oxide

One important feature of a compound is that it often looks different from the substances from which it was made. Magnesium is a grey metal and oxygen is an invisible gas, while magnesium oxide is a white powder.

The chemical formula of magnesium oxide is MgO. This tells us that the compound magnesium oxide consists of atoms of magnesium and atoms of oxygen in the ratio of 1 : 1. The composition of magnesium oxide will always be the same no matter how it is made.

There are many examples of compounds in our everyday lives.

All of the above contain one or more compounds, sometimes with water. For example, Milk of Magnesia is a **slurry** of magnesium hydroxide and water while vinegar is an **aqueous solution** of ethanoic acid.

FIG 5.8.2 Compounds found in common household products

The characteristics of compounds are summarised in Table 5.8.1. Compare them with the charactistics of mixtures given in Table 5.7.1.

Characteristics	Compound
Composition	The composition of a chemical compound is always the same.
Separation	The elements in a chemical compound can only be separated by chemical reactions.
Properties	The properties of a compound are different from the components from which it is made.
Appearance	The appearance of a compound is different from the components from which it is made.

TABLE 5.8.1

Activity 5.8.1

The effect of heat on copper

Here is what you need:

- Copper powder
- Tin lid
- Test tube × 2
- Hydrochloric acid (dilute)
- Tripod and gauze
- Heat source.

Here is what you should do:

1. Put half of the copper powder on a tin lid.
2. Place the tin lid on a tripod and gauze and heat it until no further change takes place.
3. Allow the tin lid to cool down.
4. Does the copper in the product in the tin lid look the same as the metallic copper powder? Explain your answer.
5. Add a small amount of copper powder to a test tube and add dilute hydrochloric acid to a depth of 2 cm. Gently heat the test tube and observe any changes that take place.
6. Add a small amount of product on the tin lid to a test tube and add dilute hydrochloric acid to a depth of 2 cm. Gently heat the test tube and observe any changes that take place.
7. Does the copper in the product on the tin lid have the same chemical reactions as the metallic copper powder? Explain your answer.

> **Fun fact**
>
> Some substances are mixtures of compounds. For example, petrol is a mixture of between 150 and 1000 different compounds called hydrocarbons.

Check your understanding

1. Fig 5.8.3 shows what happened when a sample of a chemical called copper(II) carbonate was weighed, heated and then reweighed.

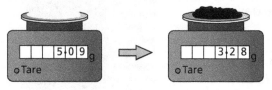

FIG 5.8.3

a) What evidence is there that a new compound has been formed?

b) Has the copper carbonate reacted with a gas from the air or has it released a gas into the air? Explain your answer.

> **Key terms**
>
> **reactants** starting materials for a chemical reaction
>
> **products** substances formed during a chemical reaction
>
> **slurry** mixture of undissolved solid and liquid
>
> **aqueous solution** a solution in water

Iron and sulfur

We are learning how to:

- describe the properties of a mixture of iron and sulfur
- describe the properties of the compound iron sulfide.

Iron and sulfur >>>

Iron is a metallic grey element and sulfur is a non-metallic yellow element.

The colour of a mixture of iron and sulfur can be any shade between grey and yellow depending on what proportion of each element is present. Mixtures do not have a fixed composition.

Iron is **magnetic** whereas sulfur is not. This difference in properties can be used to separate the elements using a magnet. Mixtures can be separated using physical processes.

FIG 5.9.1 Iron filings and sulfur powder

FIG 5.9.2 Mixtures of iron and sulfur

FIG 5.9.3 Separating iron and sulfur

FIG 5.9.4 Iron reacts with dilute acids to produce hydrogen gas

Iron reacts with dilute acids to form the gas hydrogen. When dilute hydrochloric acid is added to a mixture of iron and sulfur, the iron reacts just as it would if it was pure. The components in a mixture retain their own properties.

Iron sulfide

When a mixture of iron and sulfur is heated in a test tube they react to form the **compound** iron sulfide. We can represent this reaction by a word equation:

iron + sulfur → iron sulfide

Once the reaction has started the glass test tube continues to glow and starts to melt even when it is taken out of the flame. This is because a lot of heat is given out during the reaction. Reactions which give out heat are described as **exothermic** reactions.

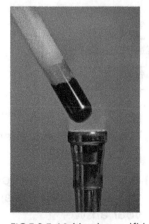

FIG 5.9.5 Making iron sulfide

Iron sulfide is a brown solid. It is very different in appearance from the grey metallic iron and yellow sulfur powder.

Although iron is magnetic, iron sulfide is not. Iron cannot be separated from sulfur in the compound iron sulfide using a magnet.

Iron reacts with dilute acids to form the gas hydrogen. When a dilute acid is added to iron sulfide the product is not hydrogen, but another gas called hydrogen sulfide. This gas is very poisonous and should only be made in a fume cupboard or near the open window of a well-ventilated laboratory.

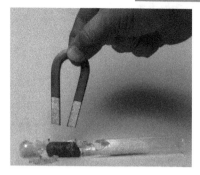

FIG 5.9.6 Iron sulfide is not magnetic

Activity 5.9.1

Comparing a mixture of iron and sulfur with iron sulfide

Here is what you need:

- A mixture of iron and sulfur
- Magnet
- Dilute hydrochloric acid
- Test tube
- Test tube holder
- Heat source.

1. Look at the mixture of iron and sulfur. Identify grey particles of iron and yellow particles of sulfur.
2. Put a mixture of iron and sulfur in a test tube to a depth of about 2 cm.
3. Place a magnet on the side of the test tube and slowly move it upwards. Explain your observations.
4. Heat the mixture of iron and sulfur in a hot flame. Once the reaction starts, remove the test tube and observe that it continues to glow. After the reaction has finished allow the test tube to cool.
5. Scrape or bang out some of the iron sulfide formed.
6. Carefully observe the iron sulfide. Can you still see individual grey and yellow particles?
7. Test whether iron sulfide is magnetic with a magnet.

Check your understanding

1. Copy and complete Table 5.9.1 to compare a mixture of iron and sulfur with iron sulfide.

Characteristics	Mixture of iron and sulfur	Iron sulfide
Composition		
Separation		
Properties		
Appearance		

TABLE 5.9.1

Key terms

magnetic attracted by a magnet

compound substance formed by the chemical combination of elements

exothermic describes a reaction that gives out heat

Separating mixtures

We are learning how to:

- explain methods of separating mixtures
- choose appropriate methods to separate mixtures
- separate mixtures using hand picking, sieving or filtration
- name the parts of the mixture obtained after filtration.

Physical separation ⟩⟩⟩

A mixture is a physical combination and can therefore be separated using physical means. How do you do this?

Activity 5.10.1

Hand picking, sieving, filtration

Here is what you need:

- Mixtures of: stones and seeds, salt and rice, soil and water
- Funnel
- Filter paper
- Glass rod
- Sieve
- Beaker
- Conical flask
- Containers
- Retort stand with clamp.

FIG 5.10.1 Different mixtures

Here is what you should do:

1. Decide if you need any apparatus to separate the stones and seeds. Try separating them. Is it easy? Why?

2. Decide if you need apparatus to separate the salt and rice. Try separating them. Which equipment did you use? Why was that a good method?

3. Decide how would you separate the soil and the water. Try separating them. Why did you choose that method? Was your method successful?

The stones are relatively large compared with the seeds, so they can be separated by hand picking each one out or by using a sieve. Salt grains are much smaller than the rice grains and so a mesh, such as a kitchen sieve, with holes that are too small for the rice to go through, can be used for separation. The method is called **sieving**. Filtration would also work with this mixture.

Filtration

The soil and water can also be separated by sieving, but in this case the 'sieve' is filter paper with tiny perforations that are much too tiny for the soil particles to go through. This method is called **filtration** or filtering.

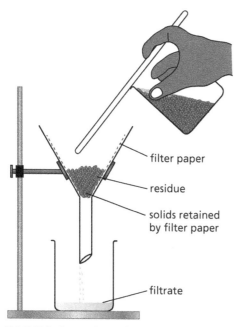

FIG 5.10.2 Separating soil and water by filtration

When the filtration process is complete, the liquid collected is called the **filtrate** and the solid left on the filter paper is called the **residue**. If the perforations of the filter paper were small enough so that not even the tiniest soil particle would go through, the filtrate would be as crystal clear as it was before the mixture was made. However, this is not likely. Filtration is only one part of the process used to purify water for human use.

Check your understanding

1. **a)** A mixture of soil and water is separated by passing through filter paper. Draw a diagram to show how this is done and label the filtrate, the filter and the residue.

   b) Why is the filtrate coloured after the separation?

5.10

Fun fact

Crystals called zeolites are used by chemists as micro-filters to remove microscopic particles from water. Zeolites have a very regular structure and the spaces between their molecules act like the perforations in filter paper to stop microscopic particles from passing through.

Key terms

sieving a method of separation where one material is smaller in size than the other

filtration a method of separation using paper with very small perforations (holes)

filtrate material that passes through filter paper

residue material that does not pass through filter paper

Evaporation and distillation

We are learning how to:

- explain methods of separating mixtures
- separate the components of a solid–liquid solution by evaporation and by distillation
- explain how the distillation apparatus works
- outline the advantages and disadvantages of distillation over evaporation.

Evaporation »»

Activity 5.11.1

Removing the liquid from a solution

Here is what you need:

- Solutions of copper sulfate and/or sodium chloride
- Evaporating dishes
- Bunsen burner
- Tripod and gauze
- Eye protection.

Here is what you should do:

1. Wear eye protection throughout.

2. Pour some of the solution into the evaporating dish.

3. Set up your apparatus as shown in Fig 5.11.1. Ensure that there is only a gentle flame.

4. Observe carefully. Do not overheat.

5. Discuss your observations.

 SAFETY

Wear eye protection when heating. Use correct procedure to light Bunsen burner. Do not touch hot apparatus and materials with bare hands. Copper sulfate is an irritant. Care is needed when handling.

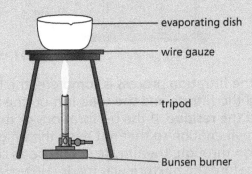

FIG 5.11.1 Apparatus set up for evaporation

Evaporation and boiling are both processes by which a liquid becomes a gas.

When an aqueous solution of crystals dissolved in water is heated, the water evaporates (and if heated strongly enough it will boil). When all of the water has escaped, only the solid is left. Because the solid has been dried very quickly it tends to dry out as a powder, not as crystals. Crystals are only produced with very slow evaporation.

Some liquids evaporate more easily than others. This property is called **volatility**. Very volatile liquids evaporate quickly at room temperature. Perfume is one example.

Activity 5.11.2

Observing distillation (1)

Your teacher will set up the apparatus as shown in Fig 5.11.2. It is called distillation apparatus and is used to separate a solution.

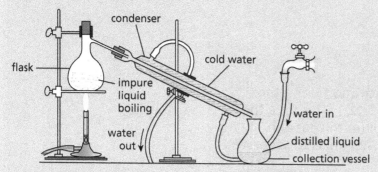

FIG 5.11.2 Distillation apparatus consists of three main parts: distillation flask, condenser, collection vessel

Here is what you should do:

1. Observe the apparatus set up by your teacher. Observe the demonstration carefully.

Distillation is used in many labs for separating or purifying components of a liquid mixture.

As the impure liquid (solution) is heated, it forms a vapour. The vapour escapes into the condenser where the cooling by the surrounding cold running water causes the vapour to condense back to a liquid. This liquid drips into the collection vessel as the distilled liquid or '**distillate**', while the solid remains as a residue in the distillation flask.

The main advantage of distillation over evaporation is that both the liquid and the solid are retrieved. For use on a large scale, however, it is time- and energy-consuming. A tremendous amount of water is wasted.

Check your understanding

1. Why is it an advantage for a perfume to be volatile?

2. In the distillation apparatus shown in Fig 5.11.2, suggest why the 'in' water to the condenser is at a lower level than the 'out' water?

3. You have now made crystals by two processes – by cooling a supersaturated solution and by evaporation. Which do you think gives the purer crystals? Why?

Fun fact

Approximately 97% of the water on Earth is salty, and therefore not drinkable. One method of obtaining fresh water is the desalination of sea water, which involves distillation.

Key terms

evaporation the process by which a liquid becomes a gas below its boiling point

volatility the ability of a liquid to evaporate

distillation method of separation that retrieves solid and liquid from a mixture

distillate distilled liquid after distillation process

Distillation of liquid mixtures

We are learning how to:

- explain methods of separating mixtures
- separate two liquids of different boiling points.

Observing distillation ▶▶▶

Activity 5.12.1

Observing distillation (2)

Your teacher will perform a demonstration using the distillation apparatus shown in Fig 5.12.1. This time there are two liquids in the flask – alcohol (in this case an alcohol called ethanol) and water. The aim is to separate these. How can this be done?

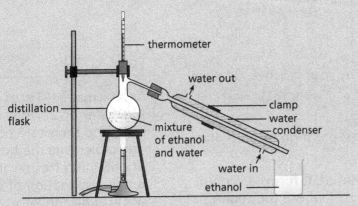

FIG 5.12.1 Distillation apparatus to separate ethanol from water

Here is what you should do:

1. Observe the demonstration carefully.

2. There is a thermometer in the flask. At what temperature do you observe vaporisation?

3. Which one of the liquids is vaporising more quickly? Why?

4. Is there condensation occurring?

5. Does your sense of smell tell you what is condensing?

6. What is happening to the temperature? Can you explain this?

7. Can you tell when all the ethanol is out of the mixture? How?

Ethanol is a volatile liquid that boils at 78 °C. Because the water temperature in the condenser is lower, the ethanol vapour condenses and liquefies. The liquid ethanol drips out into the beaker. The temperature in the flask remains at 78 °C while the ethanol boils. All the heat energy is causing the ethanol to boil. When it has all evaporated the temperature rises because the water gains energy.

The separation depends on the two liquids having widely different **boiling points.**

FIG 5.12.2 Rum is made by distilling a fermented molasses mixture. This distillery in Barbados still uses traditional copper distillation apparatus

Check your understanding

1. Complete these sentences by filling in the blank spaces:

   A liquid with a _____ boiling point is a more volatile liquid.

   A liquid with a _____ boiling point is a less volatile liquid.

2. Give two ways that you can tell how a liquid, when being heated, reaches its boiling point.

3. A mixture of water and another type of alcohol, whose boiling point is 97 °C, needs to be separated. Can distillation be used? Explain your answer.

Key term

boiling point the temperature at which a liquid boils and the liquid turns to vapour

Chromatography

We are learning how to:

- explain methods of separating mixtures
- set up a chromatography experiment
- explain how chromatography works.

Chromatography ▶▶▶

Many coloured inks are not pure colours but mixtures. Do you know how many colours make up the ink in a black or brown marker pen? It is easy to separate the ink to reveal its component colours using the method of **chromatography**.

Activity 5.13.1

Separating inks

Here is what you need:

- Sheets of chromatography or filter paper
- Beakers
- Covers
- Ethanol
- Markers of various colours
- Paper clips, pencils (or lollipop sticks or twigs)
- Scissors.

Here is what you should do:

1. Cut the filter paper into long rectangular strips so that each group member gets one strip.

2. Place a dot of marker ink (diameter about 2 mm) at a point about 1 cm from the end of your strip.

3. Place the paper clip at the other end of the filter paper.

4. Attach the paper clip to the pencil.

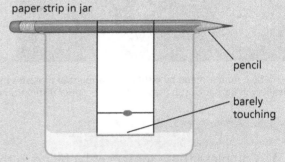

paper strip in jar

pencil

barely touching

FIG 5.13.1 Apparatus set up for chromatography

5. Pour a little ethanol into the beaker and lower your strip in so that the strip *just* touches the ethanol as shown in Fig 5.13.1.

6. Cover the beaker (to reduce evaporation of the volatile ethanol) and observe.

7. Discuss what you think will happen to the dot of ink after the paper has been left for a while in the ethanol. (Hint: ink is soluble in ethanol.)

8. You can begin to write up a laboratory report while you wait. Leave space in your observation section to paste your chromatography result when it has dried.

How chromatography works

Chromatography is a process in which a mixture, carried through a medium by a liquid (or gas), is separated into its components, because they move through the medium at different rates.

In ink chromatography, ethanol is used as a **solvent**, to dissolve the ink. The different components of the ink have different **solubilities** – they dissolve in the ethanol to different extents. This means that they are carried up the paper at different rates, and some travel further than others.

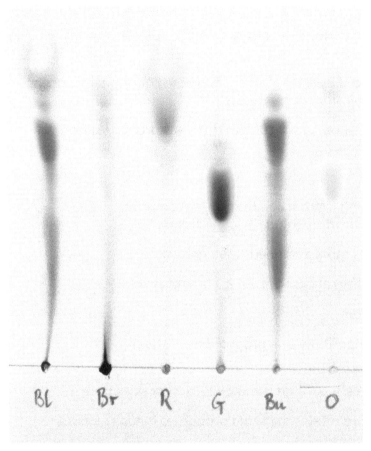

FIG 5.13.2 Results of chromatography of coloured pens and markers

Check your understanding

1. In your chromatography activity:

   a) What is the 'medium' through which the ink is carried?

   b) What is it carried by?

2. How do you think the coloured patterns in Fig 5.13.2 were produced?

Key terms

chromatography a process in which a mixture, carried through a medium by a liquid (or gas), is separated into its components, because they move through the medium at different rates

solvent liquid in which something dissolves

solubility how much something dissolves in a solvent

Review of Physical and chemical changes

- Physical properties are characteristics of a substance that can be observed without changing the nature of the substance.

- Qualitative physical properties include colour, odour and state of matter.

- Quantitative physical properties give numerical information about matter and include electrical conductivity, heat conductivity, solubility, melting and boiling point, strength, hardness, elasticity and magnetism.

- Chemical properties are the characteristics shown when a substance combines with another during a chemical change.

- When substances are heated, some undergo a physical change, some a chemical change and others no change.

- A chemical change may be evidenced by smell, colour change, coloured flame, giving out or taking in heat, sound or explosion, or effervescence.

- In a chemical reaction, a new substance is formed with different properties from the starting substances.

- Chemical changes are not usually reversible.

- Chemical changes include the rusting of iron, combustion, decomposition, food digestion, cooking, explosions and fruit ripening.

- Some chemical changes are not desirable and need to be prevented.

- Mixing and changes of state are physical changes, which are reversible.

- A mixture is a physical mix of substances.

- Mixtures do not have an exact composition. The proportions of the different components in a mixture can vary.

- The components of a mixture can be separated using physical processes.

- Each of the components in a mixture retains the same properties it has as a pure substance.

- Compounds have an exact composition. The proportion of the different elements in a compound is always the same. Those proportions determine the chemical formula of a compound.

- The elements in a compound cannot be separated using physical processes. A compound is only broken down by chemical reactions.

- A compound has a different set of properties to the elements from which it is formed.

- A mixture is a physical combination and can therefore be separated using physical means.

- There are various methods of separating a mixture, each based on the physical properties of the components.

- In filtration, a mixture of a solid and a liquid are passed through a filter. The liquid collected is the filtrate and the solid left on the filter is called the residue.

- The more volatile a liquid, the lower its boiling point. A very volatile liquid evaporates rapidly at room temperature.

- In distillation apparatus, a liquid is boiled off and then its vapour is condensed.

- Simple distillation can be used to separate the solid and liquid components of a solution, or two liquids with very different boiling points.

- The method of separation that depends on the solubility of the components is chromatography.

- Different inks dissolve at different rates in ethanol, so they travel different distances up the chromatography paper.

Review questions on Physical and chemical changes

1. State whether each of the following is a physical change or a chemical change.

   a) Meat hardens when it is grilled on a barbeque.

   b) Water turns into ice when placed in a freezer.

   c) The filament of a lamp glows white hot when electricity passes through it.

   d) Milk starts smelling bad if left out of the refrigerator overnight.

2. a) Give two examples of evidence that a chemical reaction has occurred.

   b) Why is making juice a physical change and not a chemical change?

3. State whether each of the following is pure in the scientific sense or not.

   a) Tap water

   b) Distilled water

   c) Sea water

   d) Mineral spring water

4. Which of the following are qualitative observations and which are quantitative observations that may be made in a scientific experiment?

   a) The length of a metal spring increases by 13 mm.

   b) A white precipitate is formed when two solutions are mixed.

   c) The temperature of a reaction mixture increases by 4.3 °C.

   d) Sound travels faster in water than it does in air.

5. Which of the following are physical properties and which are chemical properties?

   a) The boiling point of ethanol is 78 °C.

   b) Ammonia gas dissolves in water to give an alkaline solution.

   c) Copper(II) carbonate reacts with acids to produce carbon dioxide gas.

   d) Copper wire will conduct electricity.

6. Explain why air is a mixture and not a chemical compound in terms of:

   a) its composition

   b) how it may be separated into component

   c) its properties.

7. Explain why sodium chloride is a chemical compound and not a mixture in terms of:

   a) its composition

   b) how it may be separated into components

   c) its properties.

8. Magnesium is a silvery-grey metal. When heated in air it reacts with oxygen, burning with a bright flame to form a white powder called magnesium oxide.

   a) From the above paragraph identify:

      i) two elements

      ii) one compound

      iii) one mixture.

   b) How does the appearance of magnesium oxide differ from that of magnesium?

   c) Would you expect magnesium oxide to have the same properties as magnesium? Explain your answer.

9. State four differences between a mixture of iron and sulfur and the compound iron sulfide.

10. You have a mixture of pebbles, sand, salt and iron filings, all in water. Explain how you would separate the mixture into its five different components.

11. a) Explain how separation occurs in the process of chromatography.

    b) Describe how you would show that a sample of black ink actually contains three coloured pigments.

12. Draw a labelled diagram to show how an insoluble solid can be separated from a liquid by filtration.

Physical and chemical changes

One of the problems of recycling materials is that the different materials must be separated from each other before they can be used again.

Scrap metal dealers are able to separate iron and steel from other metals because they are attracted by a magnet.

A recycling centre has a large amount of an unusual mixture of waste materials which result from a manufacturing process. The mixture consists of small polythene beads, iron filings, aluminium dust and tiny lead shot.

You have been asked to use your knowledge of separating techniques to devise a method of separating these four materials so they can be reused.

FIG 5.SIP.1 Separating iron and steel from other metals

1. You are going to work in groups of 3 or 4 to devise methods of separating these materials. The tasks are:

   • To review the content of the unit and particularly methods of separation and how they relate to the properties of the materials concerned.

   • To carry out research into other methods of separation which are not covered in this unit.

   • To consider the properties of the materials in the mixture.

   • To devise methods for separating the mixture.

   • To make up a mixture of known proportions for testing purposes.

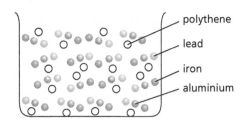

FIG 5.SIP.2 Mixture to be separated

   • To test your method and evaluate the results.

   • To modify your method in the light of the test results.

   • To compile a report including a PowerPoint presentation in which you should explain how you went about solving the problem and how you arrived at your recommendations. You should illustrate your report by taking pictures at different stages during testing.

   a) Look back through this unit, and particularly the different methods of separation described. Notice that the method used to separate a mixture depends on the properties of the components.

   b) Carry out some research into other methods of separation which are not covered in the unit. Here's one that might give you some ideas.

   It is possible to separate particles of gold from grains of sand because gold is much denser. If a mixture is swirled with water and the water is poured away, the gold tends to be left behind.

c) Make a list of some of the properties of the materials in the mixture that might help you to think of ways of separating them. For example, you might make a table of their densities.

Material	Density (g/cm³)
Polythene	0.92
Iron	7.87
aluminium	2.70
Lead	11.34

TABLE 5.SIP.1

FIG 5.SIP.3 Panning for gold

It might help you to know that the density of water is 1 g/cm³. Anything which has a density less than this will float in water while anything that has a density more than this will sink.

d) Use the information you have gathered to devise ways of separating the mixture. You will not be able to separate the four components at the same time. Rather you will have to use a sequence of three different methods; each method will separate one material from those that remain. The order in which you apply the methods may be important.

e) Make up a sample mixture containing 25 g each of the four components. You can use this to find out how well your separation methods are working.

f) Carry out a trial separation using your methods. Don't forget to take some pictures to illustrate your report. At the end of the trial, dry and weigh each of the four separated components. How close to 25 g of each component did you obtain?

You should modify one or more of your methods on the basis of:

- How easy the methods are to carry out

- How good a separation was obtained

Carry out a second trial incorporating any modifications you think will improve the process.

g) Prepare a PowerPoint presentation in which you describe what you did in order to bring about the separation of the four components. You should illustrate your account with photographs.

Show the data you obtained in the form of separations of the test sample and relate this to how well or how badly particular methods of separation worked.

Unit 6: Forces and motion

We are learning how to:

- describe that forces can cause motion
- explain what forces can do.

Forces and motion ≫

Forces

A force is a push or a pull on one object by another. A force can change the motion or the shape of an object.

FIG 6.1.1 A bowler exerts a push force on a cricket ball

The force exerted by a bowler causes the cricket ball to speed up.

The force exerted by the parachute on the air causes the aircraft to slow down. The parachute resists the flow of air.

When a vehicle collides with an object the forces exerted on it change its shape.

Forces as vectors

Forces can be described as vector quantities because they have both size or magnitude, and direction. They are often represented by arrows. The direction of the arrow gives the direction of the force while the length of the arrow gives an indication of its size.

FIG 6.1.2 A parachute exerts a pull force on an aircraft

FIG 6.1.3 Forces may alter the shape of an object

Balanced forces

Balanced forces are equal in size and act in opposite directions.

FIG 6.1.4 In a tug of war both teams exert a pull force

While both teams are pulling equally hard the forces are balanced and neither team moves. When one team weakens, the forces become unbalanced and that team is pulled forwards.

Motion

Motion is about how objects move, or how they are moved in relation to other objects.

In a race all of the runners start running at the same time. Some will cover the distance more quickly than others. How can we express speed in terms of distance and time?

Sometimes we talk about the velocity of a moving object. Is this just another word for speed or does it mean something different in science?

What is acceleration? Is it related to velocity in some way?

> **Fun fact**
>
> Some objects can exert a force on others without actually touching them. For example a magnet attracts a piece of iron.

FIG 6.1.5 What is the relationship between speed, time and distance?

FIG 6.1.6 Some sports cars can accelerate from 0 to 100 km/h in only a few seconds

Scalars and vectors

We are learning how to:

- explain the difference between scalar and vector quantities
- give examples of scalar and vector quantities.

Scalar and vector quantities

The quantities used in science can be conveniently placed into two groups:

- **Scalar** quantities are quantities that have size or **magnitude** but no direction.

- **Vector** quantities have both magnitude and direction.

Table 6.2.1 gives some common examples of each group.

Notice that some quantities are related. For example:

- Displacement is distance in a particular direction.

- Velocity is speed in a particular direction.

Also notice that a quantity derived from a vector is itself a vector. For example:

- Acceleration is the rate of change of velocity. Because velocity is a vector quantity, acceleration is also a vector quantity.

Examples of scalar quantities	Examples of vector quantities
Mass Volume Density Distance Speed Energy Time Temperature	Force Displacement Velocity Acceleration

TABLE 6.2.1

Representing forces

Forces are vector quantities because they have both magnitude and direction. Forces are measured in units called **newtons** (N).

We can represent forces in a diagram using arrows.

5 N   3 N   1 N

FIG 6.2.1 Forces of different magnitude acting in the same direction

The forces represented by the arrows in Fig 6.2.1 are different magnitudes but are acting in the same direction. Notice that the length of the line is proportional to the size of the force. The line representing a force of 5 newtons (5 N) is five times longer than the line representing a force of 1 N.

3 N   3 N   3 N   3 N

FIG 6.2.2 Forces of the same magnitude acting in different directions

The forces represented by the arrows in Fig 6.2.3 are **equal** in magnitude but they are acting in different directions.

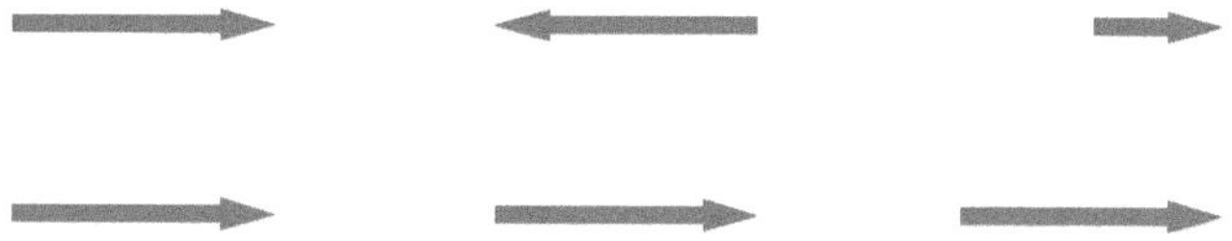

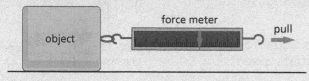

FIG 6.2.3 Forces that are equal and forces that are not equal

For two forces to be equal, they must be both equal in magnitude and acting in the same direction.

Activity 6.2.1

Measuring forces

Here is what you need:

- Force meter (0–10 N)
- Small objects with loops of string attached
- Ruler.

Here is what you should do:

1. Place an object on the table in front of you.

2. Attach a force meter to the object and pull the object in a straight line.

FIG 6.2.4

3. Read the value on the force meter as you pull the object.

4. Using a scale in which 1 cm represents 1 N, draw the force. Remember that you must represent both the magnitude of the force and its direction.

5. Repeat steps 1–4 on some different objects. Pull each object in a different direction.

> **Fun fact**
>
> Forces can be added just like numbers but to do this we need to draw geometric constructions representing both their magnitude and directions. These are called force diagrams.

Check your understanding

1. The arrows in Fig 6.2.5 represent six forces, A–F.

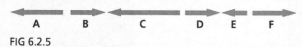

FIG 6.2.5

a) Which arrow represents the largest force?

b) Which arrow represents the smallest force?

c) Which arrows represent a pair of forces equal in magnitude but opposite in direction?

d) Which arrows represent a pair of equal forces?

Key terms

scalar quantity that has only magnitude

magnitude another word for size

vector quantity that has both magnitude and direction

newton the SI unit of force; $1 \text{ N} = 1 \text{ kg m/s}^2$

Balanced forces

We are learning how to:

- give examples of situations involving balanced forces
- explain the effect of balanced forces.

Balanced forces ⟫

When two forces acting on an object are equal in magnitude and opposite in direction the forces are said to be balanced and they will not alter the motion of the object. One force cancels the effect of the other.

The magnitude of the pull of the rope is exactly equal to the **weight** of the window cleaner and acts in the opposite direction. This keeps the window cleaner in the same position.

The magnitude of the buoyancy force is equal to the weight of the boat so it sits at the same level in the water. You will learn more about buoyancy later in this unit.

A person exerts a force on the ground by virtue of his or her weight. There must be an equal and opposite force preventing them from sinking into the ground. We call this the **reaction force**.

When the driving force of a car engine is equal to the **resistance forces** the car moves at a constant speed.

Pull of the rope

Weight of the window cleaner

FIG 6.3.1 The window cleaner stays in one position

Buoyancy force

Weight of the boat

FIG 6.3.2 The boat lies in the water

Driving force of the car   Resistance force

FIG 6.3.4 The car travels at a constant speed

Reaction force of the ground

Weight of the student

FIG 6.3.3 The student does not sink into the ground

Balancing forces

Here is what you need:

- String long enough to overhang both sides of your desk
- Weights
- A bag of sand of unknown weight.

Here is what you should do:

1. Tie a loop at either end of the string.

2. Tie a knot at the centre of the string.

3. Place the string on the table so the knot is at the centre. Mark the position of the knot by placing a pencil or pen next to it.

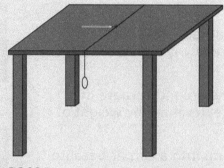

4. Hang weights of equal amounts on the two loops and observe whether the knot moves.

FIG 6.3.5

5. Hang weights of different amounts on the two loops and observe whether the knot moves.

6. Summarise your observations in terms of balanced and unbalanced forces.

7. Use the effect of balanced forces to find the weight of the bag of sand.

> **Fun fact**
>
> An object moving in a circle always has an unbalanced force acting on it, therefore the motion of a planet about the Sun, or a moon about a planet, is not an example of a balanced force.

1. Explain each of the following in terms of balanced forces.

   a) When a book is placed on a table it remains there although it has weight.

   b) A hot air balloon is able to hang in the air the same height above the ground.

   c) In a tug of war neither team moves forwards or backwards.

   d) A cyclist is travelling at a constant speed.

Key terms

weight force exerted by an object downwards due to the pull of gravity

reaction force equal force acting in the opposite direction to another force

resistance forces friction and air resistance which resist the motion of a moving object

Unbalanced forces

We are learning how to:

- give examples of situations involving unbalanced forces
- explain the effect of unbalanced forces.

Unbalanced forces

When two forces are not balanced they are said to be unbalanced. **Unbalanced forces** may alter the motion of an object. For example:

- If an object is still, unbalanced forces will cause it to start moving.

- If an object is moving, unbalanced forces may cause it to move more quickly, more slowly or in a different direction.

A rocket is able to rise up from the ground because the rocket motors produce a force greater than the weight of the rocket.

A car is able to slow down coming into a corner because the braking force is greater than the driving force of the engine.

A parachutist falls slowly to the ground because his or her weight is just a little more than the **air resistance** created by the parachute.

A moving steel ball changes direction when it experiences the pulling force of a magnet. It is an unbalanced force because there are no other forces acting.

Thrust of the rocket engines

Weight of the rocket

FIG 6.4.1 The rocket motors develop lots of thrust

Driving force of the car

Braking force

FIG 6.4.2 The car slows down coming into a corner

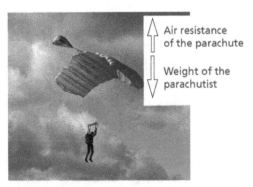

Air resistance of the parachute

Weight of the parachutist

FIG 6.4.3 A parachutist falls slowly to the ground

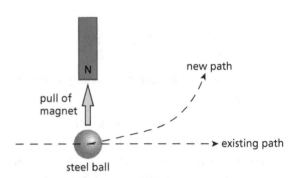

N

pull of magnet

new path

existing path

steel ball

FIG 6.4.4 Magnets attract iron and steel

Activity 6.4.1

Unbalanced forces

Here is what you need:

- Beaker
- Ping-pong ball
- Tennis ball
- Stone
- Feather.

Here is what you should do:

1. Half-fill a beaker with water.

2. Place a ping-pong ball on the surface of the water.

3. Push the ping-pong ball down under the water with the end of your pencil and then release it.

4. Explain what happens in terms of unbalanced forces.

5. Throw a tennis ball into the air and observe what happens.

6. Explain in terms of unbalanced forces why the tennis ball first rises into the air and then falls back to the ground.

7. Hold a stone and a feather at the same height above the ground and release them at the same time.

8. Observe what happens and explain your observations in terms of unbalanced forces.

> **Fun fact**
>
> Unbalanced forces acting on an object in motion may cause it to accelerate or decelerate.

Check your understanding

1. Explain each of the following in terms of unbalanced forces.

   a) A stone falls to the bottom of a beaker of water.

   b) A car goes faster when the driver pushes down on the accelerator.

   c) A helium balloon rises into the air.

   d) When a batsman hits a cricket ball it changes direction.

Key terms

unbalanced forces any forces that are not balanced

air resistance force acting to resist the movement of an object through the air

Friction

We are learning how to:

- explain the force of friction
- measure the friction force on a moving object.

Friction »»

Whenever one object slides over another, a force works against this movement. This force is called **friction**.

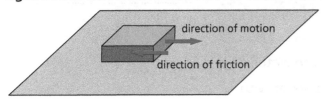

FIG 6.5.1 Friction is a contact force

The direction of friction is always in the opposite direction to the motion of an object. Friction exists only while an object is in motion.

If you examine the surface of an object that feels perfectly smooth to the touch using a powerful microscope, you will find that it is far from smooth. The surface is really an irregular pattern of high points and low points.

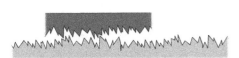

FIG 6.5.2 Surfaces moving over each other magnified

As an object moves, the high points on the object and the high points on the surface push against each other, so they act to slow the object down.

Activity 6.5.1

Investigating friction

Here is what you need:

- Piece of wood fitted with a hook
- Force meter
- Weights
- Sheet of sandpaper.

Here is what you should do:

1. Place a piece of wood with a hook in it on the table, and attach it to a force meter.

2. Gently pull on the force meter and pull the block at a steady speed.

3. Record the reading on the force meter.

4. Put a weight on top of the block, and pull the block at a steady speed.

5. Record the reading on the force meter.

6. Put a large sheet of sandpaper on top of the table so that you can pull the block over it.

7. Pull the block over the sandpaper at a steady speed.

8. Record the reading on the force meter.

9. What two factors affect the size of the friction force? Explain your answer.

Energy is needed to overcome friction. This energy is wasted, in the sense that it is not used to do useful work but is lost as heat. Friction also causes surfaces to wear each other away.

Friction cannot be completely eliminated; however, it can often be reduced by keeping surfaces apart, so that they can pass over each other more easily.

Check your understanding

1. Friction is important in many sports. State whether each of the following increases or decreases friction.

   a) A gymnast rubs chalk on her hands before she grabs the horizontal bars.

   b) A skier waxes the underneath of his skis before skiing down a mountain.

   c) A sprinter wears spiked running shoes.

   d) An ice skater wears skates that have a very thin blade.

Fun fact

On a cold day, rubbing your hands together makes them warm, because the energy needed to overcome friction between your hands is converted to heat.

Key term

friction a force that opposes the movement of one surface over another

Water resistance and air resistance

We are learning how to:

- explain water resistance and air resistance
- describe the effect that resistance has on motion.

Water resistance and air resistance ≫

Friction is not restricted to movement between solid surfaces. It also exists when a solid moves through a liquid or gas, and between moving layers of liquids and gases. In this case, we call the forces 'water resistance' and 'air resistance'.

The amount of friction experienced as an object passes through a liquid or gas depends very much on its shape.

Most fish are long and flat. They experience very little resistance, so they can move quickly through the water.

Parachutes are spread out. They experience a large amount of resistance, so the parachutist can descend to the ground slowly.

Resistance makes it more difficult for an object to move through water or air. Energy is wasted as heat.

Modern cars, aircraft and boats have shapes that reduce resistance to a minimum. This is called **streamlining**. Moving a streamlined shape wastes less fuel and makes the vehicle more efficient.

FIG 6.6.1 Fish move quickly through water

FIG 6.6.2 Parachutists move more slowly through the air

Activity 6.6.1

Investigating resistance and shape

Here is what you need:

- Modelling clay
- Tall clear container, such as a large measuring cylinder
- Cooking oil
- Rod – long enough to reach the bottom of the container
- Stopwatch.

FIG 6.6.3 Modern cars are streamlined

Here is what you should do:

1. Take a piece of modelling clay. Roll it into a ball and then divide it into four pieces of approximately equal mass.

2. Make four different shapes out of the four pieces.

3. Pour cooking oil into a tall clear container, until the level is about 2 cm from the top.

4. Hold one of the shapes so that it is just touching the surface of the cooking oil. Release the shape and start timing.

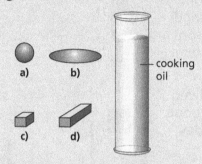

FIG 6.6.4

5. Stop timing when the shape touches the bottom of the measuring cylinder.

6. Remove the shape from the oil by pushing a long rod into the shape and lifting it out.

7. Repeat steps 4–6 for each shape.

8. Record your results in a table.

9. Draw a bar chart to illustrate your results

Check your understanding

1.

FIG 6.6.6

Say how the labelled features in Fig 6.6.6 help the cyclist to go faster.

Fun fact

Racing cars have become more and more streamlined in their design as engineers have tried to reduce air resistance so their cars will go faster.

FIG 6.6.5

Key terms

friction a force that opposes the movement of one surface over another

water resistance the friction experienced by an object moving through water

air resistance the friction experienced by an object moving through air

streamlined shaped in such a way as to minimise the effects of water resistance or air resistance

Floating and sinking

We are learning how to:

- explain why some objects float while others sink
- describe the forces involved in floating and sinking.

The Archimedes Principle

If you drop an iron nail a few grams in mass into a beaker of water it will instantly **sink** to the bottom. However, a steel ship with a mass of many thousands of tonnes is able to **float**. To understand the reason for this, we need to consider the forces that act on a body submerged in water.

Archimedes was an ancient Greek scientist who lived over 2 000 years ago. He discovered something about floating and sinking which is called the **Archimedes Principle**.

When an object is partially or wholly immersed in water, it receives a **buoyancy** force or **upthrust** equal to the weight of water it displaces. This force acts in the opposite direction to the weight of the object.

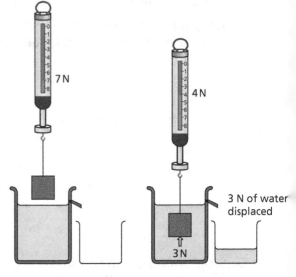

FIG 6.7.1 Buoyancy force or upthrust

Floating and sinking

The **density** of cork is only about one-quarter that of water. If a cube of cork is placed in water only about one-quarter of it needs to submerge in order to displace its own weight of water.

The density of iron is nearly eight times that of water. Even when a cube of iron is totally submerged in water it only displaces about one-eighth of its weight so it sinks.

If the steel contained in the hull of a ship was made into a solid ball and placed in the water it would sink. The weight of water displaced would be far less than the weight of the ball.

When the steel is made into a ship's hull it displaces a sufficient weight of water to float. The buoyancy force is equal to the weight of the hull.

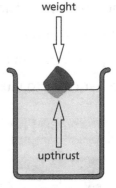

6.7.2 Cork is less dense than water

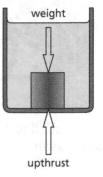

FIG 6.7.3 Iron is more dense than water

Activity 6.7.1

Investigating whether materials float or sink

Here is what you need:

- Dish of water
- Objects made from different materials.

Here is what you should do:

1. Draw a table with two columns. At the top of the first column write 'Floats' and at the top of the second write 'Sinks'.

2. Select an object and identify the material from which it is made.

3. Place the object in a bowl of water and determine whether it floats or sinks.

4. Write the name of the material in the appropriate column of your table.

5. Repeat steps 2–4 for all of the objects.

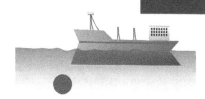

FIG 6.7.4 The hull of a ship displaces its own weight of water

The density of water is 1 g/cm³.

- Substances that have a density less than 1 g/cm³ float because they can displace their own weight of water by being only partially submerged.

- Substances that have a density more than 1 g/cm³ will sink because they cannot displace their own weight of water even when fully submerged.

FIG 6.7.5 Hot air is less dense than cold air

Archimedes Principle applies to any fluid, i.e. any liquid or gas. We can use it to explain why objects like hot air balloons can float in the air.

Key terms

sink fall to the bottom of a liquid or gas

float sit on the surface of a liquid or gas

Archimedes Principle explains why objects float or sink in terms of forces acting on them

buoyancy force acting upwards on an object in a liquid or gas

upthrust another name for buoyancy

density the mass per unit volume of a substance

Check your understanding

1. Table 6.7.1 shows the densities of two liquids, water and ethanol, and a solid, paraffin wax.

Substance	Density (g/cm³)
Ethanol	0.79
Paraffin wax	0.88
Water	1.00

TABLE 6.7.1

In terms of the forces acting on it, explain why a ball of paraffin wax floats on water but sinks in ethanol.

Speed

We are learning how to:

- investigate motion of a body
- measure time and distance.

High speed or low speed? »

When you travel from one place to another you may move quickly or you may move slowly.

FIG 6.8.1 Runners in a race move very quickly (they move at a high speed)

FIG 6.8.2 People out for a stroll move slowly (they move at a low speed)

Speed is a measure of how quickly somebody or something travels the distance between two points. One way to measure speed is to count the number of metres that a person travels each second. This is written as metres per second, **m/s**, or m s^{-1}.

A runner might travel 10 metres each second, which is a speed of 10 m/s, while somebody out for a stroll might only travel 1 metre each second, which is a speed of 1 m/s.

Activity 6.8.1

Moving slowly and moving quickly

Here is what you need:

- Measuring tape
- Two pegs
- Stopwatch.

Here is what you should do:

1. Place a peg in the ground.
2. Measure a distance of 100 m from the peg using a measuring tape.

3. Place a second peg in the ground.

4. Time how long it takes you to walk slowly from one peg to the other.

5. Next, time how long it takes you to run quickly from one peg to the other.

6. How much longer did you take to walk the distance than you took to run the same distance?

Cars usually move much more quickly than people. You could measure the number of metres a car travels each second but it is usually more convenient to measure the number of kilometres it travels each hour. This is written as kilometres per hour, **km/h** or km h⁻¹.

Cars are fitted with instruments called speedometers, which indicate their speed.

FIG 6.8.3 Cars are fitted with instruments called speedometers, which indicate their speed

Check your understanding

1. Look at the speedometer in Fig 6.8.3.

   a) At what speed is this car travelling?

   b) How far would it go if it travelled at this speed for 1 hour?

   c) Is the car moving slowing or quickly?

2. Make a list of things that move quickly and a list of things that move slowly. Give your answer in the form of a table.

Things that move quickly	Things that move slowly

Fun fact

The cheetah is the world's fastest land animal.

It can travel at speeds up to 120 km/h over short distances.

Key terms

speed a measure of how quickly somebody or something travels the distance between two points

m/s (or m s⁻¹) number of metres travelled in a second, a unit of speed

km/h (or km h⁻¹) number of kilometres travelled in an hour, a unit of speed

Relationship between speed, distance and time

We are learning how to:

- investigate motion of a body
- calculate speed.

Calculating speed »»

You can express the relationship between speed, distance and time in the form of a simple mathematical equation.

$$\text{speed (m/s)} = \frac{\text{distance (m)}}{\text{time (s)}}$$

The unit of speed depends on the units in which distance and time are given. If distance was expressed in kilometres and time in hours, then the unit of speed would be kilometres per hour or km/h.

When a person moves from one place to another their speed seldom remains the same. They might speed up for some parts of their journey and slow down for others.

FIG 6.9.1 In a 100 m race, the runner starts running slowly, increases their speed to a maximum at between 50 and 60 metres into the race, and then starts to slow down as they get tired

FIG 6.9.2 A journey by car is not at a constant speed. For some parts of the journey the driver can travel quickly but for others he or she must slow down or stop

When you divide the distance travelled by the time taken, you are finding their **average speed** over the journey.

Worked example 6.9.1

A motorist travels between two towns that are 200 km apart. The journey takes 2.5 hours. What is the average speed over the journey?

Solution

You do not know whether the motorist is travelling at a constant speed for the whole of the journey or if he or she goes faster at some times and slower at others. But since you know both the total distance and the total time taken you can find the average speed.

$$\text{average speed} = \frac{200 \text{ km}}{2.5 \text{ h}} = 80 \text{ km/h}$$

Sometimes it may be necessary to rearrange the equation. For example, you may wish to find the distance travelled by an object when you know its speed and the time it travels.

Multiplying both sides of the equation by time gives:

$$\text{speed} \times \text{time} = \frac{\text{distance}}{\text{time}} \times \text{time}$$

$$\text{distance} = \text{speed} \times \text{time}$$

Worked example 6.9.2

A ball travels at a constant speed of 3 m/s for 5 s. How far does the ball travel?

Solution

In this question you are given the speed and the time so you need the form of the equation that lets you work out the distance.

$$\text{distance} = 3 \text{ m/s} \times 5 \text{ s} = 15 \text{ m}$$

Check your understanding

1. **a)** An aircraft travels 7150 km from London to Port of Spain in a time of 11 hours. What was the average speed of the aircraft for the journey?

   b) Make time the subject of the equation relating speed, distance and time.

   c) Calculate the time for the same journey if the speed of the aircraft was increased to 700 km/h.

Fun fact

FIG 6.9.3 Usain Bolt

As of August 2015, Usain Bolt held the world record for running 100 m in a time of 9.58 s. His average speed during the race was therefore $\frac{100}{9.58} = 10.44$ m/s.

Key term

average speed speed calculated by dividing distance travelled by time taken

Distance–time graphs

We are learning how to:

- investigate motion of a body
- draw and interpret distance–time graphs.

Distance–time graphs ⟩⟩

You can represent the movement of an object as a graph by plotting the distance it has travelled at different times on its journey.

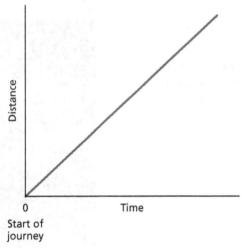

FIG 6.10.1 Distance–time graph of object moving at constant speed

The shape of a **distance–time graph** depends on how the object moves. If an object moves at a constant speed the distance–time graph will be a straight line.

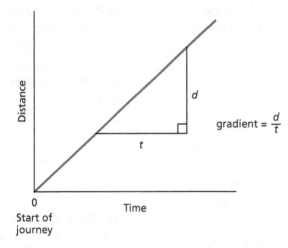

FIG 6.10.2 Gradient of a distance–time graph

The speed of an object moving at constant speed is given by the slope or gradient of the distance–time graph.

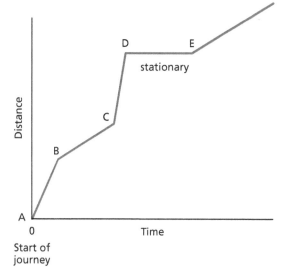

FIG 6.10.3 The speed is not the same between points A and B as it is between points B and C, points C and D and after point E

If an object changes speed as it moves, the graph will no longer be a straight line. If part of the graph is horizontal (as it is between points D and E), this means that the object is stationary for that period of time and the distance from the start of the journey remains unchanged.

Check your understanding

1. On a distance–time graph:

   a) What information is given by the gradient of the graph?

   b) How can you tell that an object is stationary?

Fun fact

If a distance–time graph is a curve, the speed at any point can be found by drawing a straight line or tangent to the curve at that point, and calculating its slope.

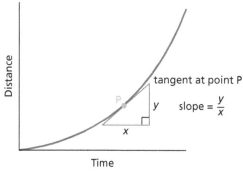

tangent at point P

$slope = \frac{y}{x}$

FIG 6.10.4

Key term

distance–time graph graph that shows how the distance travelled varies with time

Displacement and velocity

We are learning how to:

- investigate motion of a body
- account for displacement and velocity both in terms of magnitude and direction.

Magnitude and direction

Distance and speed are both examples of scalar quantities. They have size or magnitude but they do not have a particular direction. For example, you might say that a ball travels at 0.5 m/s. You have not specified in which direction, therefore 0.5 m/s represents its speed and in 1 s it will travel a distance of 0.5 m.

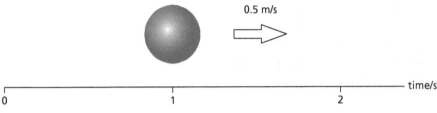

FIG 6.11.1

The ball in Fig 6.11.1 is travelling from left to right. The motion of the ball now has both magnitude and direction so you can say that the ball travels at a velocity of 0.5 m/s from left to right, and in 1 s it will be displaced to the right by a distance of 0.5 m.

Displacement and **velocity** are examples of vector quantities. They have both magnitude and direction.

- Displacement is the distance travelled in a particular direction.
- Velocity is the speed in a particular direction.

Activity 6.11.1

Measuring your velocity

You should work with a partner for this activity.

Here is what you need:

- Measuring tape
- Two pegs
- Stopwatch.

Here is what you should do:

1. Place a peg in the ground.

2. Measure a distance of 100 m from the peg using a measuring tape in a direction given by your teacher.

3. Place a second peg in the ground.

4. Time how long it takes your partner to walk from the first peg to the second peg in the given direction.

5. Calculate your partner's velocity when they walked.

6. Next, time how long it takes your partner to run from the first peg to the second peg in the given direction.

7. Calculate your partner's velocity when they ran.

8. When was your partner's velocity greater?

Fun fact

FIG 6.11.3 Rockets need to have a high velocity

When a rocket is launched it travels upwards. You talk about its velocity rather than its speed because it travels in a particular direction. In order to overcome the pull of the Earth's gravity, a rocket must achieve a velocity of 11.2 km/s. This is called the escape velocity. If it does not achieve this velocity it will fall back down to the ground when it runs out of fuel.

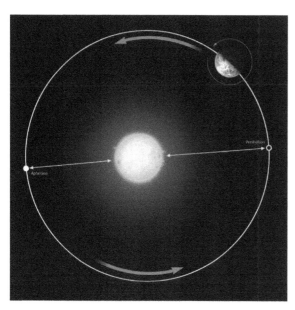

FIG 6.11.2 The Earth is in orbit around the Sun

The Earth is in orbit around the Sun at a constant speed of 30 km/s. You cannot say that the Earth has a constant velocity of 30 km/s because its direction is continually changing. To have a constant velocity, an object would need to travel both at the same speed and in the same direction.

Check your understanding

1. a) Explain the difference between speed and velocity.

   b) The Moon orbits the Earth at a constant speed of about 1 km/s. Why is it incorrect to say that the Moon orbits the Earth at a constant velocity?

Key terms

displacement the distance travelled in a particular direction

velocity the speed in a particular direction

Acceleration

We are learning how to:

- investigate motion of a body
- explain acceleration as a change in velocity.

Changing velocity ⟫

An object accelerates when the forces acting on it are not balanced. This may result in the velocity of the object increasing or decreasing. A decrease in velocity is called deceleration.

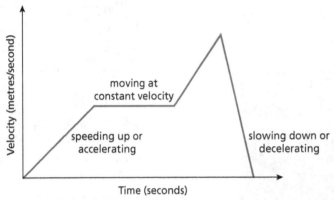

FIG 6.12.1 A velocity–time graph shows how the velocity of an object changes over time

The shape of the graph indicates how the velocity of the object is changing.

- When the graph is sloping upwards the velocity of the object is increasing; the object is accelerating.

- When the graph is horizontal the object is moving with a constant velocity.

- When the graph is sloping downwards the velocity of the object is decreasing; the object is decelerating.

Acceleration is the rate of change of velocity with respect to time. You can express the relationship between acceleration, velocity and time in the form of a simple mathematical equation.

$$\text{acceleration (m/s}^2) = \frac{\text{change in velocity (m/s)}}{\text{time (s)}}$$

When velocity is expressed in metres per second, and time in seconds, the unit of acceleration is the metre per second per second, m/s^2 (or m s^{-2}).

The acceleration of a moving object over a period of time is given by the gradient of the velocity–time graph over that period.

Key terms

acceleration rate of change of velocity with respect to time

m/s² (or m s⁻²) units of acceleration

Worked example 6.12.1

Fig 6.12.2 shows the velocity of an object over a period of 10 s.

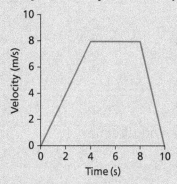

FIG 6.12.2

What is the acceleration of the object between these times?

a) 0–4 s **b)** 4–8 s **c)** 8–10 s

Solution

The equation you use is:

$$\text{acceleration} = \frac{\text{change in velocity}}{\text{time}}$$

a) $\text{acceleration} = \frac{8 - 0}{4} = 2 \text{ m/s}^2$

b) $\text{acceleration} = \frac{8 - 8}{4} = 0 \text{ m/s}^2$ so the object is at constant velocity

c) $\text{acceleration} = \frac{0 - 8}{2} = -4 \text{ m/s}^2$; negative acceleration is deceleration

Worked example 6.12.2

How long will it take a car to slow from 30 m/s to 12 m/s if it decelerates at a rate of 1.5 m/s²?

Solution

$$\text{time} = \frac{\text{change in velocity}}{\text{acceleration}} = \frac{30 - 12}{1.5} = \frac{18}{1.5} = 12 \text{ s}$$

Check your understanding

1. In order for the space shuttle *Discovery* to reach the minimum altitude required to orbit the Earth, it had to accelerate from 0 to a velocity of 8 000 m/s in 8.5 minutes.

   a) How many seconds are there in 8.5 minutes?

   b) Calculate the acceleration over this period in m/s², assuming it is constant.

Fun fact

FIG 6.12.3

A top fuel dragster can accelerate from 0 to 576 m/s in 0.8 seconds. That is an acceleration of 720 m/s².

Review of Forces and motion

- Forces can:
  - change the motion of an object
  - change the direction of a moving object
  - change the shape of an object.

- A scalar quantity has magnitude only.

- A vector quantity has both magnitude and direction.

- Force is a scalar quantity.

- Forces can be represented as arrows.

- An arrow indicates the direction of a force and its length the magnitude of the force.

- Balanced forces are equal in magnitude and opposite in direction.

- Balanced forces do not alter the motion of an object.

- Unbalanced forces are forces which are not balanced.

- Unbalanced forces can start a stationary object moving or alter the speed and/or direction of a moving object.

- Friction is a force on surfaces when they move over each other.

- Friction when an object moves in water or in air is called water resistance or air resistance.

- Streamlining involves shaping objects so that they have as little resistance as possible.

- When an object is placed in water it receives a buoyancy force or upthrust equal to the weight of water displaced. This is known as the Archimedes Principle.

- The Archimedes Principle can be applied to any liquid or gas.

- If an object displaces its own weight of water when partially submerged it will float.

- If an object cannot displace its own weight of water even when totally submerged it will sink.

- Speed is the rate at which an object moves from one point to another. It is measured in metres per second, m/s, or kilometres per hour, km/h. The equation for speed is:

$$\text{speed (m/s)} = \frac{\text{distance (m)}}{\text{time (s)}}$$

- A distance–time graph shows distance on the vertical or y-axis and time on the horizontal or x-axis. The speed of an object is given by the slope or gradient of a distance–time graph.

- Displacement is the distance an object moves in a particular direction and is measured in the same units, for example metres, m.

- Velocity is the speed an object travels in a particular direction and is measured in the same units, for example metres/second, m/s. The equation for velocity is:

$$velocity \ (m/s) = \frac{displacement \ (m)}{time \ (s)}$$

- Acceleration is the rate at which velocity changes over time. It is a result of unbalanced forces acting on an object. Acceleration is measured in metres per second per second, m/s^2. The equation for acceleration is:

$$acceleration \ (m/s^2) = \frac{change \ in \ velocity \ (m/s)}{time \ (s)}$$

Review questions on Forces and motion

1. Fig 6.RQ.1 shows the directions of three forces acting on a racing car.

FIG 6.RQ.1

   a) In which direction would an applied force:
   i) slow the car down?
   ii) hold the car on the road?
   iii) speed the car up?
   b) Say whether each force is a push or a pull.

2. Fig 6.RQ.2 shows a book with two forces acting on it.

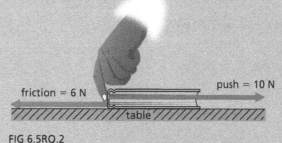

FIG 6.5RQ.2

   a) What unit does 'N' represent?
   b) What is the significance of the arrow that represents a force?
   c) What is the resultant force on the book?
   d) Are these examples of contact forces or non-contact forces?

3. Describe whether each of the following is a scalar quantity or a vector quantity. Give your answer in the form of a table.

| temperature | force | velocity | density |
| energy | length | acceleration | speed |

4. Using a scale of 1 cm represents 1 N draw arrows to represent the following.

a) A 4 N force acting to the right.

b) A 6 N force acting upwards.

c) A 5 N force acting to the left.

5. Marie is a downhill skier.

a) Why does Marie not sink into the snow when she is wearing skis, but she does sink when she is wearing her ordinary shoes?

b) Why does Marie rub wax on the underneath of her skis?

c) Why does Marie crouch down when she is skiing?

FIG 6.RQ.3

6. A 'speed camera' is able to detect the speed of a car. The lines on the road are exactly 2 m apart. The camera takes two photographs, the second exactly 0.5 s after the first.

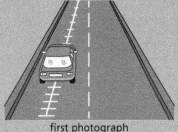

first photograph   second photograph

FIG 6.RQ.4

a) The speed limit on this road is 60 km/h. What is this speed in m/s?

b) How far did the car travel between the first and second pictures?

c) Calculate the speed of the car and decide if it was exceeding the speed limit.

7. Fig 6.RQ.5 shows how the velocity of a car changes over a short journey.

a) During which period, A, B or C, is the car:

  i) slowing down?

  ii) speeding up?

  iii) travelling at a constant velocity?

b) What is the fastest velocity at which the car travels? Give this in m/s and in km/h.

c) What is the acceleration of the car in the first 50 s of the journey?

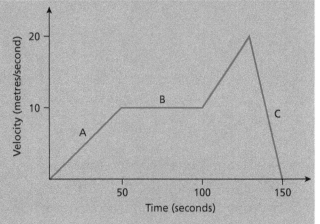

FIG 6.RQ.5

8. a) Explain why less force is needed to lift an object when it is partially submerged in water compared to when it is on the ground.

b) Draw a diagram of a hot air balloon and show the relative sizes and directions of the forces acting on it when it is hovering at a constant height above the ground.

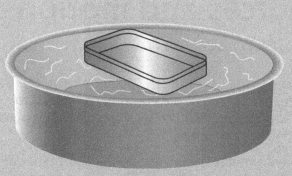

FiG 6.RQ.6

c) In an experiment an aluminium tray, used to store food in a takeaway restaurant, was placed on a bowl of water and floated.

The tray was removed from the bowl, squashed into a ball and placed back on the surface of the water.

i) Predict what will happen when the aluminium ball is released.

ii) Explain your prediction.

Forces and motion

The density of water is 1 000 kg/m³. The relative density of a liquid is its density compared to that of water. Relative density is also sometimes called specific gravity.

A hydrometer is a device that is used to measure the relative density of liquids. This can be used to identify the liquid.

Hydrometers look a little like fishing floats. The have:

- ballast, to keep them upright in the liquid
- a float to allow them to float in the liquid
- a graduated stem from which values can be read.

As a scientist who understands about floating and sinking your task is to make a hydrometer and use it to investigate whether temperature affects the density of a liquid.

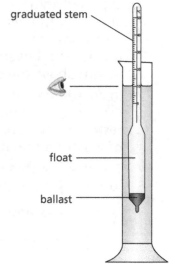

graduated stem

float

ballast

FIG 6.SIP.1 A hydrometer

1. You are going to work in groups of 3 or 4 to make a working hydrometer. The tasks are:

   - To review the work on floating and sinking, and the Archimedes Principle, to make sure you understand how a hydrometer works.
   - To design and make a hydrometer and then to calibrate your hydrometer.
   - To test your hydrometer by measuring the relative densities of some liquids whose relative densities are already known. Modify its design if necessary.
   - Use your hydrometer to investigate whether temperature affects the density of a liquid chosen by your teacher.
   - To compile a report including a PowerPoint presentation in which you explain how you built your hydrometer and tested it. Your report should also include the results of the investigation carried out with your hydrometer. You should illustrate your report by taking pictures at different times during the investigation.

2. Look back through the lesson on floating and sinking. When an object is placed in a liquid think about how the volume of liquid displaced is related to its density. The denser a liquid, the smaller the volume of liquid that needs to be displaced in order to equal the weight of an object.

3. How are you going to build a hydrometer? Here are some simple examples which might give you some ideas.

   Hydrometers don't have to be made of glass. Plastic tubes work just as well. Have a look around your home and see what you can find. Drinking straws are good, or how about the empty barrel or shell of a ball point pen? You can block any unwanted holes with modelling clay.

   You need to find some ballast, like small nails or lead shot. Use enough to make your hydrometer stand upright in a liquid but leave plenty of tube above the surface.

Experiment with whatever materials are available until you are happy you have something that will work. Take some pictures of your instrument at different stages of manufacture.

4. To calibrate your hydrometer you will need two liquids whose relative densities are known. Your teacher will help you with this. Table 6.SIP.1 has examples of liquids you might use.

Choose two liquids whose relative densities are as far apart as possible. Mark the level of your hydrometer tube at the surface when placed in both liquids in turn. Using a ruler, draw a scale on your tube using the marks to guide you.

For example, if the relative densities of the two liquids were 900 and 1100 and the marks were 2 cm apart then 2 cm in length would be equivalent to 1100 – 900 = 200 difference in relative density.

5. Now is the time to test your hydrometer by measuring the relative densities of some liquids whose values are already known. Your teacher will help you with this.

How accurate were your readings and how easy was your hydrometer to read? As a result of this test, modify the design of your hydrometer in a way which you think will make it better.

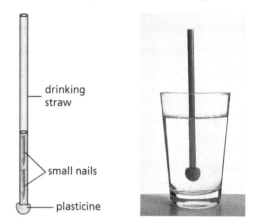

FIG 6.SIP.2 Some ideas for building hydrometers

drinking straw

small nails

plasticine

d) Early scientists often made the apparatus and equipment they used to carry out experiments and that is what you have done. You are going to use what you have made to investigate if the relative density of a liquid, provided by your teacher, changes with temperature.

Liquid	Relative density at room temperature
Hexane	655
Ethanol	789
Propan-1-ol	804
Sea water	1025
Dichloromethane	1326
1,2,4-Trichlorobenzene	1454

TABLE 6.SIP.1

To do this you need to measure the relative density of the liquid at several different temperatures. Take some pictures of different stages of your investigation.

e) The final task is to prepare a PowerPoint presentation in which you explain what you have done. This will be in two parts. The first part should deal with the making of your hydrometer and the second with the investigation into the effect of temperature on relative density. You should illustrate your account with photographs or drawings.

Discuss any pattern that you might have observed between the relative density of a liquid and its temperature.

Unit 7: Respiration and gas exchange

We are learning how to:
- understand the structure of the respiratory system
- distinguish between breathing, gas exchange and respiration.

Respiration and gas exchange »

The respiratory system is responsible for providing the body with oxygen and removing the waste gas, carbon dioxide. These gases pass around the body in the circulatory system.

Structure of the respiratory system

Air is taken into the body through the nose and sometimes through the mouth. Both of these connect to the trachea, which carries air down into the lungs. You can see that a person is breathing because their chest moves up and down, even when they are asleep.

Exchanging gases

Oxygen gas forms about one-fifth of the air around us. It is essential for life. Without oxygen you would die. Carbon dioxide is a waste gas produced by the body. This must be removed before it damages the body.

In the lungs these gases are exchanged. Oxygen passes from the air into the body, and carbon dioxide passes in the opposite direction.

Breathing

Breathing is one of those actions that your body does automatically without you having to think about it. You started breathing the moment you were born and you will continue to breathe until the moment you die.

Breathing, however, requires muscular action, which alters the shape and size of the thoracic cavity that contains the lungs. The change in shape and size produces changes in the pressure inside the cavity compared to the external pressure of the atmosphere.

Cell respiration

The term 'respiration' is used to describe both the action of breathing in and out and the chemical process that takes place in the cells of the body to release energy.

These processes are linked in that breathing in and out is the way in which the body is able to supply and remove the gases associated with respiration in the cells.

Breathing rate

Your breathing rate is the number of times you breathe in and out each minute. As you get older your breathing rate tends to get slower.

FIG 7.1.1 Breathing rate increases when you are active

Because breathing is associated with providing the body with oxygen in order to produce energy, it is not surprising that breathing rate is linked to level of activity. You breathe more frequently when you are active than when you are at rest.

Energy from food

Food provides the body with the energy released during cell respiration.

FIG 7.1.2 Food gives us energy

The amount of energy provided by a food is called its calorific value. Different foods provide different amounts of energy.

Structure of the respiratory system

We are learning how to:

• outline the basic structure of the respiratory system
• identify the different parts of the respiratory system.

The respiratory system ≫

The respiratory system is concerned with breathing. Oxygen is needed by the body for respiration and during this process carbon dioxide is produced. These gases are exchanged in the **lungs**. Water vapour is also lost from the body in exhaled air.

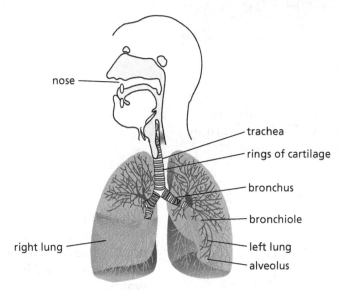

nose

trachea

rings of cartilage

bronchus

bronchiole

right lung

left lung

alveolus

FIG 7.2.1 Structure of the respiratory system

There are four parts to the respiratory system.

• The **nose**, where air is inhaled and exhaled.

• The **trachea**, which carries air into and out of the chest.

• The bronchi (singular **bronchus**), which connect the trachea to each of the lungs.

• The two lungs, which are found on either side of the thoracic (chest) cavity, and are protected by the ribcage.

When you breathe in, air passes through the nose or mouth into the trachea. The trachea then divides into two bronchi, one serving each lung. Each bronchus divides many times to form smaller tubes called **bronchioles**. Each bronchiole ends in a collection of tiny air bags called **alveoli**.

Activity 7.2.1

Feeling the rings of cartilage in the trachea

You do not need any equipment or materials for this activity.

Here is what you should do:

1. Place your thumb pointing upwards on your throat just above your ribcage.

2. Gently push your thumb up and down your throat.

3. You should be able to feel some rings of cartilage. They are like the rings of wire inside the hose of a vacuum cleaner.

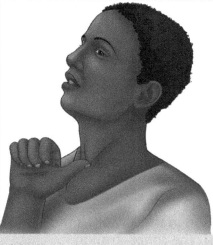

FIG 7.2.2

4. Cartilage is a stiff material that is more flexible than bone. Why do you think the trachea has rings of cartilage?

The capacity of the lungs increases as you grow up. The lung capacity of a young adult man is about 5.8 dm³ while for a woman it is 4.2 dm³. When you reach around 30 years old, the capacity starts to decrease.

When a person is 50 years old their lung capacity will only be about half of what it was in their youth. A reduced lung capacity means that less oxygen enters the body so less energy is obtained from respiration in the body cells. This explains why growing older is associated with shortage of breath and decreased endurance. Older people are also more susceptible to respiratory disorders.

Check your understanding

1. Arrange the following in the order that air passes through when you breathe in.

   alveolus   bronchiole   bronchus   nose   trachea

Key terms

lungs organs of the body found on either side of the thoracic (chest) cavity and protected by the ribcage

nose where air is inhaled and exhaled

trachea carries air into and out of the chest

bronchus connects the trachea to a lung

bronchioles smaller tubes that connect to a bronchus

alveoli small air sacs found at the end of bronchioles

Gas exchange in the lungs

We are learning how to:

- outline the basic structure of the respiratory system
- explain the exchange of gases in the lungs.

Gas exchange »»

Air is a mixture of gases. The main constituents are nitrogen (78%) and oxygen (21%). Air also contains a very small concentration (0.04%) of carbon dioxide.

When air enters an alveolus (one of the alveoli) these gases can pass through the alveoli walls and the walls of the blood capillaries that surround them.

Gases **diffuse** in both directions between the alveoli and the blood capillaries. The overall effect of **gaseous exchange** is determined by the relative **concentration** of each gas.

- **Oxygen** is in higher concentration in the alveoli as this is air that has just been inhaled. The concentration of oxygen in the blood is low as this is deoxygenated blood. So there will be a net movement of oxygen into the blood from the air.

- **Carbon dioxide** is produced by the body during respiration. Deoxygenated blood contains a much higher concentration of carbon dioxide than air. The result is a net movement of carbon dioxide from the blood into the air.

- Nitrogen diffuses into and out of blood in the same way as any other gas but, as it plays no part in cell respiration, the concentration in the blood remains the same. There is no net movement of nitrogen into or out of the blood.

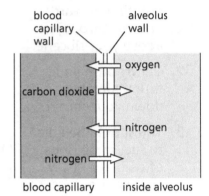

FIG 7.3.1 Diffusion of gases in the alveoli

Fun fact

The concentration of nitrogen dissolved in blood is not significant on land but it can affect a diver.

FIG 7.3.2

The deeper a diver goes, the more gas dissolves in their blood. If a diver rises to the surface too quickly, nitrogen gas will come out of solution and form bubbles at the joints. This condition is called 'the bends' and it is very painful.

Activity 7.3.1

To compare the concentration of carbon dioxide in inhaled and exhaled air

Here is what you need:

- Two boiling tubes with bungs
- Tubing
- T-piece
- Limewater
- Two stands and clamps.

Here is what you should do:

1. Set up the apparatus as shown in Fig 7.3.3. It is important that the tubes are the correct length and orientation.

2. Pour limewater into each boiling tube until it is about half full.

3. Slowly breathe in and out through the mouthpiece.

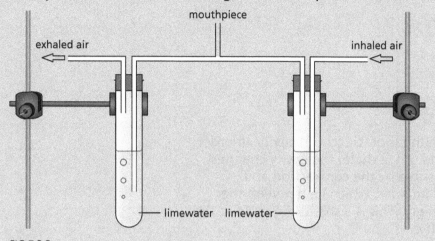

FIG 7.3.3

4. Limewater turns cloudy in the presence of carbon dioxide. In which boiling tube did the limewater turn cloudy first?

5. What can you deduce about the concentration of carbon dioxide in exhaled air compared to inhaled air?

Water is also produced during cell respiration. Much of this is lost through the lungs.

Exhaled air is always saturated in water vapour. If you breathe on a cold window or a mirror the water vapour condenses, forming a layer of tiny water droplets.

FIG 7.3.4 Water vapour condenses on a mirror

Key terms

diffuse when gases move between two places

gaseous exchange exchange of gases from one place to another

concentration amount of gas present in a particular volume (of air)

oxygen a gas taken into the cells for respiration

carbon dioxide a gas produced by the body during respiration

Check your understanding

1. Table 7.3.1 gives some information about the composition of inhaled and exhaled air.

	% of oxygen	% of carbon dioxide
Inhaled air	21	0.04
Exhaled air	16	4

TABLE 7.3.1

a) What fraction of the oxygen in inhaled air is absorbed into the body?

b) By how much is the concentration of carbon dioxide in exhaled air greater than the concentration in inhaled air?

How you breathe

We are learning how to:

- distinguish between breathing and respiration in humans
- explain breathing in terms of changes in the volume of the chest cavity.

Breathing 〉〉

The lungs are situated in the thoracic (chest) cavity. In order to breathe in (**inhale**) and out (**exhale**) the cavity changes volume. This is made possible by the contraction and relaxation of **intercostal muscles**, which lie between the ribs, and also the **diaphragm**. This is a sheet of muscle at the bottom of the chest cavity.

Inhaling

When the intercostal muscles contract, they lift the ribs upwards and outwards. At the same time the diaphragm contracts and flattens. The result of this is that the volume of the chest cavity increases. The pressure of air decreases. The air pressure in the lungs is now less than atmospheric pressure so air is forced into the lungs through the trachea.

The trachea has rings of cartilage along its length. These prevent the trachea from collapsing when the pressure in the lungs falls, or it would be impossible for air to enter the lungs.

Exhaling

This is the opposite process to inhaling.

The intercostal muscles relax, allowing the ribs to drop downwards and inwards. At the same time, the diaphragm muscle relaxes and it curves upwards. The result is a reduction in the volume of the chest cavity. The air pressure in the cavity is now greater than atmospheric pressure so air is forced out of the lungs through the trachea.

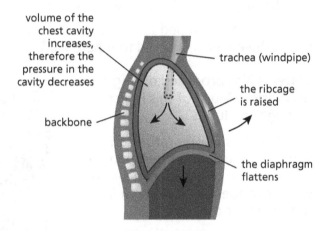

volume of the chest cavity increases, therefore the pressure in the cavity decreases

trachea (windpipe)

the ribcage is raised

backbone

the diaphragm flattens

FIG 7.4.1 Inhaling

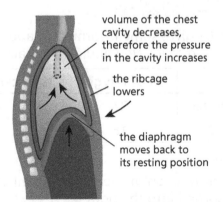

volume of the chest cavity decreases, therefore the pressure in the cavity increases

the ribcage lowers

the diaphragm moves back to its resting position

FIG 7.4.2 Exhaling

Activity 7.4.1

Breathing in and out

Your teacher will lead this activity.

Here is what you need:

* Bell jar and bung
* Rubber sheet
* Y-tube
* Two balloons
* Elastic band.

Here is what you should do:

1. Connect balloons to the two forks of a Y-tube using elastic bands.

2. Place the end of the Y-tube through a bung in the bell jar from the inside.

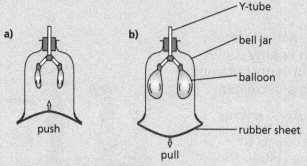

FIG 7.4.3

3. Draw a rubber sheet over the bottom of the bell jar and hold it in place with elastic bands.

4. Push the rubber sheet upwards as in Fig 7.4.3a) and observe any change in the size of the balloons.

5. Pull the rubber sheet downwards as in Fig 7.4.3b) and observe any change in the size of the balloons.

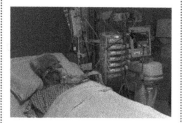
Check your understanding

1. **a)** Draw a table to show how the volume of the chest cavity and the pressure inside it change as a person breathes in and out.

   b) Name the muscles that contract and relax:

   i) between the ribs

   ii) at the bottom of the ribcage.

Key terms

inhale breathe in

exhale breathe out

intercostal muscles muscles that lie between the ribs

diaphragm a sheet of muscle at the bottom of the chest cavity

Respiration in cells

We are learning how to:

- distinguish between breathing and respiration in humans
- describe the process of respiration that takes place in the cells of the body.

Respiration »»

The term 'respiration' is a term which must be used accurately in science.

- Respiration is not breathing in and out. Breathing in and out should be called ventilation or inhalation and exhalation.

- Respiration is the process by which energy is released in cells by the reaction between food and oxygen. This process is also known as **internal**, **cell** or **tissue respiration**.

All body cells require energy to power the many different chemical processes that go on in them. This energy comes from the chemical reaction between the nutrients obtained from the digestion of food, such as glucose and oxygen:

food + oxygen → carbon dioxide + water + energy

Carbon dioxide and water are the waste products of respiration. These are carried away from the cells in the blood. Respiration produces the carbon dioxide that is expelled from the lungs. Excess water is lost from the body by breathing, sweating and as urine.

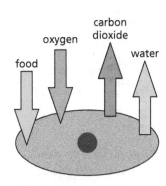

FIG 7.5.1 Respiration in a cell

Activity 7.5.1

Making a model of a cell

Make a model cell and use it to show how substances move into and out of the cell.

Here is what you need:

- Materials for making a model, for example plastic bottles, small balls
- Tools for cutting and shaping, for example scissors, craft knife.

Here is what you should do:

1. Build a model of a simple cell to show the nucleus and cytoplasm.
2. Show the movement of substances into and out of the cell in some way.

Within the cytoplasm of all cells – plant and animal – there are structures called mitochondria (singular **mitochondrion**).

Mitochondria are sometimes called the 'powerhouses' of the cell because it is here that cell respiration takes place.

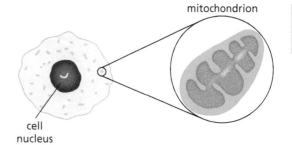

FIG 7.5.2 Mitochondria are in animal and plant cells

The number of mitochondria in different cells varies. Cells that require a continuous supply of energy have the highest numbers of mitochondria.

Heart muscle continually contracts and relaxes throughout our lives without a rest. Heart cells have a high number of mitochondria.

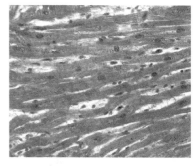

FIG 7.5.3 Heart muscle cells

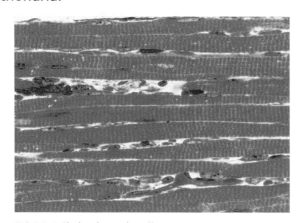

FIG 7.5.4 Skeletal muscle cells

Skeletal muscle is sometimes called voluntary muscle as it only contracts and relaxes when you decide to move. It requires less energy. These cells have a lower number of mitochondria than heart muscle cells.

> **Fun fact**
>
> Mitochondria vary in size and in number from one to over a thousand per cell. However, all mitochondria have the same basic structure.

Check your understanding

1. This equation represents the combustion of a fuel such as natural gas:

   fuel + oxygen → carbon dioxide + water + energy

   a) In what ways is this reaction similar to cell respiration?

   b) In what ways is this reaction different from cell respiration?

Key terms

internal (cell, tissue) respiration the process by which energy is released in cells due to the reaction between food and oxygen

mitochondrion structure in cell where cell respiration takes place

Breathing rate

We are learning how to:

- distinguish between breathing and respiration in humans
- measure breathing rate.

Breathing rate »»

Your **breathing rate** is how many times you breathe in and out each minute.

FIG 7.6.1 Breathing rate is low when you are relaxed or when you are asleep

FIG 7.6.2 Breathing rate increases when you are more active

Activity 7.6.1

Measuring breathing rate

If possible, work with a partner for this activity.

When somebody knows their breathing rate is being measured they become self-conscious and breathe more quickly. A doctor measures the breathing rate of a patient by observing them without the patient realising what is happening.

Here is what you need:

- Stopwatch.

Here is what you should do:

1. Sit opposite your partner so you can watch each other.
2. Over a period of 10 minutes, time the breathing rate of your partner five times but do not tell them when you are doing it.
3. Start timing and then count how many breaths they take in 1 minute.
4. Write down the number in a table like the one here.

	1st time	2nd time	3rd time	4th time	5th time
Number of breaths in 1 minute					

5. What was your partner's average breathing rate?

6. Collect the results of the students in the class. Use them to create a comparison table of the average breathing rate of the boys against girls. Your table must be constructed accurately – title, headings – rows and columns, all units in headings and the content recorded accurately.

7. What do you notice?

Are all the breathing rates the same?

Are there noticeable trends?

Are the breathing rates of the boys similar to those of the girls?

Are there athletes in your class?

Are the breathing rates of the athletes the same as those of the non-athletes?

8. Interview the teachers in the Physical Education Department to find out how exercise affects breathing rate.

Breathing rate changes with age.

Age in years	Number of breaths per minute
Up to 1	30–60
1–3	24–40
3–6	22–34
6–12	18–30
12–18+	12–20

TABLE 7.6.1

As you grow older, the number of breaths you take each minute decreases. Breathing rate is always given as a range because people are all built a little bit differently.

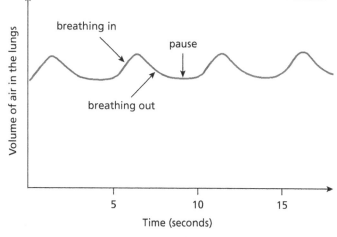

FIG 7.6.3 A typical pattern for breathing rate for a person at rest – each cycle of breathing in and breathing out is followed by a brief pause

Check your understanding

1. a) Estimate the breathing rate of the person in Fig 7.6.3 from the information given.

   b) If Fig 7.6.3 represents the breathing rate at rest, suggest what age group the person belongs to.

Key term

breathing rate how many times you breathe in and out each minute

Exercise and breathing rate

We are learning how to:

- distinguish between breathing and respiration in humans
- explain why exercise affects breathing rate.

Effect of exercise on breathing rate ›››

Oxygen is needed by the body to make energy in the body.
The only way the body can obtain oxygen is by breathing.

FIG 7.7.1 When the body is at rest, normal breathing provides enough oxygen for the small amount of energy needed

FIG 7.7.2 When you exercise you must breathe more often and more deeply to get the extra oxygen needed to provide more energy

Activity 7.7.1

Further investigation of breathing rates

Here is what you need:

- Stopwatch.

In Activity 7.6.1, you worked in pairs to measure your breathing rate and then your partner's breathing rate. The class collected information which can be used for further activities.

Review the information that was collected. Are all the breathing rates the same? Are there noticeable trends? Are the breathing rates of the boys similar to those of the girls? Are there athletes in your class? Are the breathing rates of the athletes the same as those of the non-athletes?

Let us see if the rates change with exercise. Continue working in pairs.

1. Do some form of mild exercise, e.g. walking? Do this for 5 minutes.

2. At the end of the 5 minutes, sit down and allow your partner to measure your breathing rate in the next minute. Record the number.

3. Now do some form of stronger exercise, e.g. running? Do this for 5 minutes.

4. At the end of the 5 minutes, sit down and allow your partner to measure your breathing rate in the next minute. Record the number.

5. Design and create a table to show the trends that you have noticed. Your table must be constructed accurately – title, headings, rows and columns, all units in headings and the content recorded accurately.

6. What is the effect of exercise on the breathing rate?

If a person takes strenuous exercise, such as an athlete running very quickly, even rapid breathing cannot supply the body with enough oxygen for respiration. The body becomes short of oxygen. A period of recovery is required after exercise.

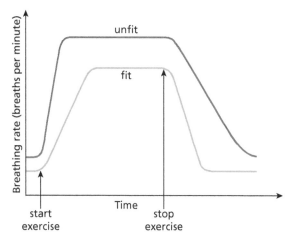

FIG 7.7.3 The time a person needs to recover depends on how fit their body is (the fitter they are the quicker they will recover and breathe normally again)

> **Fun fact**
>
> When a person exercises they may get a build-up of a substance called lactic acid in their muscles. The build-up of lactic acid causes pain in the muscles (a stitch). After they have finished exercising their body needs oxygen to break down the lactic acid. The amount of oxygen needed to do this is called the oxygen debt. This is why people carry on breathing hard even after they have finished exercising.

Check your understanding

1. a) What happens to a person's breathing rate when they exercise?

   b) Explain why it does this.

Energy from food

We are learning how to:

- relate the contents of food to the energy it supplies the body
- measure the amount of energy released by a sample of food.

Nutrients in food >>

You will already know from the work you did in Unit 4 that the nutrients in food can be placed into different groups and these include carbohydrates, fats and proteins. The body obtains most of the energy it needs from carbohydrates and from fats. Energy is measured in **joules** and kilojoules (1 000 J = 1 kJ).

CHOCOLATE MUFFIN (1 pc) 1715 kJ

FRENCH FRIES (1 cup) 1590 kJ

SODA POP (1 bottle) 565 kJ

POTATO CHIPS (15 chips) 670 kJ

MULTIGRAIN BREAD (1 slice) 460 kJ

VEGETABLE PIZZA (1 slice) 1495 kJ

FIG 7.8.1 Foods that provide lots of energy

Foods that are rich in carbohydrates (which include sugars) and fats therefore provide the body with most energy. However, the amount of energy provided by a food also depends on how it is cooked. For example, a jacket potato will not provide the same amount of energy as an equal mass of French fries or potato chips.

The amount of energy obtained from food is roughly:

- 17 kJ per gram of carbohydrates and per gram of proteins
- 37 kJ per gram of fats.

It is possible to use these figures to obtain an estimate of the amount of energy in a food. Table 7.8.1 shows the energy available from 100 g of potato chips.

Fun fact

In a food calorimeter the amount of energy released by a food is accurately measured by burning the food in pure oxygen.

FIG 7.8.2 Food calorimeter

Nutrient	Amount in 100 g portion (g)	Amount of energy per gram (kJ)	Total amount of energy (kJ)
Carbohydrates	36.6	17	36.6 × 17 = 622.2
Fats	14.5	37	14.5 × 37 = 536.5
Proteins	4.1	17	4.1 × 17 = 69.7
		Total	1228.4 kJ

TABLE 7.8.1

This shows that 100 g of potato chips contains about 1228 kJ of energy.

Activity 7.8.1

Investigating the energy released by burning a food

Here is what you will need:

- Boiling tube
- Measuring cylinder 25 cm³ or 50 cm³
- Mounting needle
- Piece of food
- Thermometer
- Stand and clamp
- Bunsen burner or similar
- Balance.

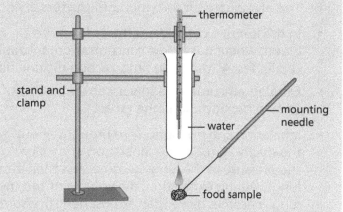

FIG 7.8.3 Foods release energy when they burn

Here is what you should do:

1. Measure 25 cm³ of water into a boiling tube.

2. Support the boiling tube on a stand and clamp and suspend a thermometer in the water.

3. Record the initial temperature of the water.

4. Weigh a piece of food, such as a peanut, and record its mass.

5. Place the food on a mounting pin. Start the piece of food burning by holding it in the flame of a Bunsen burner and then quickly move it under the boiling tube.

6. Keep the burning food under the boiling tube until it stops burning. Record the final temperature of the water.

7. Use your results to calculate the calorific value of the food using the method below.

The amount of energy released by a food can be estimated by burning a known mass, and using the energy given out to heat a known volume of water.

It needs 4.2 joules of energy to raise the temperature of 1 g of water by 1 °C. You can use this information to calculate the energy released by the food in the activity:

energy released (J) = 4.2 × mass of water (g) × temperature rise (°C)

Check your understanding

1. Which components of food provide the body with most of its energy?

2. A 50 g portion of a breakfast cereal contains 24 g of carbohydrates, 7 g of proteins and 1.8 g of fats. Estimate the amount of energy this provides the body.

Key term

....

joule unit of energy

Review of Respiration and gas exchange

- The respiratory system is the system of the body that is concerned with breathing.

- Air passes into the body through the nose and mouth and then through the trachea, bronchi, bronchioles and finally the alveoli in the lungs.

- The alveoli are air sacs found in clusters at the ends of the bronchioles.

- In the lungs, oxygen gas diffuses from the air, where it is in high concentration, into the blood, where it is in low concentration, through the alveoli walls and into the blood capillaries. At the same time, carbon dioxide diffuses out of the blood and into the air.

- Exhaled air contains a lower concentration of oxygen and a higher concentration of carbon dioxide than inhaled air.

- Inhalation and exhalation are brought about by the action of the intercostal muscles and the diaphragm. During inhalation, the volume of the chest cavity increases and therefore the pressure in the lungs becomes less than atmospheric pressure. Air is forced into the lungs. During exhalation, the volume of the chest cavity decreases and the pressure in the lungs becomes more than atmospheric pressure. Air is forced out of the lungs.

- Cell respiration is the process by which cells use nutrients from food and oxygen to provide energy:

  food + oxygen → carbon dioxide + water + energy

- Cell respiration takes place in structures called mitochondria, which are found in the cytoplasm of a cell. Heart muscle requires a lot of energy so heart muscle cells have a large number of mitochondria. Skeletal muscle needs less energy so skeletal muscle cells have fewer mitochondria.

- Breathing rate is the number of times a person breathes in and out each minute. The rate decreases as you grow to adulthood. A typical value for an adult male is between 12 and 20 breaths per minute.

- Breathing rate increases when the body exercises because it requires more oxygen for cells to produce more energy.

- The fitter a person is, the more quickly their body will recover after exercise and start breathing normally again.

- Food provides the body with energy.

- Different foods provide different amounts of energy.

- The body obtains most of the energy it needs from carbohydrates and fats.

- A gram of carbohydrate provides the body with about 17 kJ of energy.

- A gram of fat provides the body with about 37 kJ of energy.

Review questions on Respiration and gas exchange

1. Copy and complete the following sentences.

    a) When you breathe in your lungs fill with _____ .

    b) In the lungs _____ passes into the blood and _____ passes out.

    c) The body also loses excess _____ when you breathe out.

    d) The number of times a person breathes in and out each minute is called their _____ .

2. Fig 7.RQ.2 shows some parts of the body associated with respiration.

    a) Name the parts labelled A, B, C and D.

    b) Identify the other organ, shown in the chest cavity but not labelled.

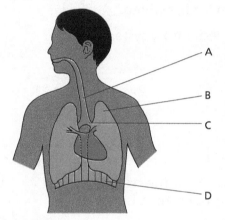

FIG 7.RQ.1

3. The apparatus in Fig 7.RQ.2 is used to investigate gaseous exchange during breathing.

    a) Limewater is used to test for which gas?

    b) What happens to the limewater if the gas is present?

    c) Which of the tubes in Fig 7.RQ.2 would show the presence of this gas first?

    d) What does the result of this investigation demonstrate?

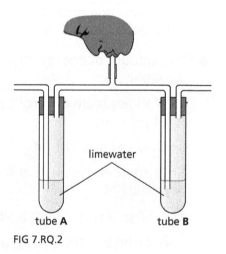

limewater

tube **A**   tube **B**

FIG 7.RQ.2

4. Amanda ran for 5 minutes. Fig 7.RQ.3 shows how her breathing rate changed after she stopped running and rested.

a) i) What was Amanda's breathing rate at the point when she stopped running?

ii) What is Amanda's normal breathing rate when she is at rest?

iii) How long did it take for Amanda's breathing rate to return to normal after she had stopped running?

b) Explain why running increases the breathing rate.

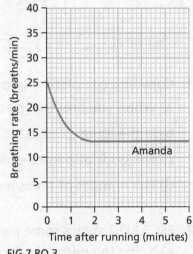

FIG 7.RQ.3

5. Fig 7.RQ.4 shows a model that can be used to demonstrate part of the mechanism for breathing.

a) Which parts of the respiratory system are represented by:

i) the bell jar?

ii) the glass tube?

iii) the balloons?

iv) the rubber sheet?

b) Explain why the volume of the balloons changes when the rubber sheet is moved up and down.

c) Suggest one reason why this is not a good model to show breathing.

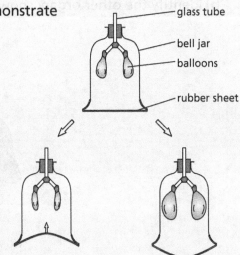

FIG 7.RQ.4

6. The amount of energy obtained from food is roughly:

• 17 kJ per gram of carbohydrates and per gram of proteins

• 37 kJ per gram of fats.

Here are some nutritional facts from a bag of potato crisps.

a) In 50 g of crisps what is the mass of:

i) carbohydrates   ii) proteins   iii) fats?

b) Estimate the amount of energy a person would obtain from eating 50 g of the crisps.

FIG 7.RQ.5

7. The graph shows how the volume of air in the lungs changes when breathing in different ways.

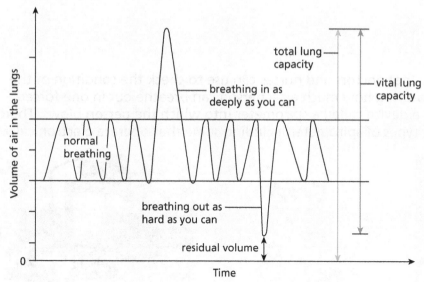

FIG 7.RQ.6

Use the graph to state whether each of the following statements is true or false.

a) When you breathe out as hard as you can your lungs contain no air.

b) Normal breathing only changes part of the air in the lungs.

c) The vital lung capacity is the maximum amount of air that the lungs can contain.

d) A deep breath takes in twice as much air as a normal breath.

e) An experiment to measure lung capacity by breathing in deeply and blowing air into a calibrated vessel actually measures vital lung capacity and not total lung capacity.

8. A sample of 0.30 g of a breakfast cereal was placed in a flame to ignite it. Once it was burning it was held under a boiling tube containing 25 cm³ of water at 25.0 °C, held by a stand and clamp. The temperature of the water after the cereal was completely burnt was 34.1 °C.

a) What was the temperature increase of the water of the boiling tube?

b) It requires 4.2 J of energy to raise the temperature of 1 g of water by 1 °C. How much energy was needed to raise the water in the boiling tube to its final temperature?

c) Use the results given to calculate the amount of energy in 100 g of the breakfast cereal.

d) The actual amount of energy in 100 g of the cereal is significantly more than the answer to **c)**. Suggest two features of the method used that would affect the accuracy of the result.

Respiration and gas exchange

Spirometry is a simple test that doctors and nurses can use to check the condition of the lungs. They do this by measuring how much air a person can breathe out in one forced breath. This is done using a device called a spirometer into which the person blows. There are a number of different types of spirometer including some that operate electronically.

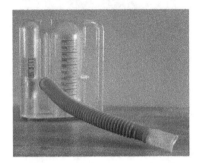  

FIG 7.SIP.1 Different kinds of spirometers

Spirometers are useful because they give an indication that the lungs are not functioning correctly. This indicates to the doctor that further investigations and tests are necessary.

The problem is that spirometers can be expensive to buy. You have been approached by the World Health Organization to design a spirometer that can be made out of simple materials. This can then be built and used by doctors in poor countries where there is insufficient money to buy such equipment.

1. You are going to work in groups of 3 or 4 to design and build a spirometer. The tasks are:

   - To find out what you can about spirometers and how they work.
   - To design and build a prototype spirometer.
   - To calibrate your spirometer.
   - To test your spirometer to find out such things as how accurate it is and how easy it is to use.
   - To modify your design on the basis of what you found out during testing.
   - To write instructions, including diagrams and photographs, for making a spirometer which can be used by doctors in other countries.

   a) Use whatever reference material is available to find out about spirometers and how they work. If you have access to the internet you can type 'spirometer' into a search engine. You will also find spirometers for sale on different sites.

   Your local medical centre may have a spirometer. Ask one of the nurses if he/she would demonstrate how it works. Take some pictures with your phone.

FIG 7.SIP.2 How long is the balloon after one breath out?

b) Now you need to consider some different designs. Here are some ideas to get you started.

You could experiment with balloons. Can the amount of exhaled air in one forced breath be related to the increase in the length or diameter of a balloon?

Is the height to which you can blow a ping-pong ball related to the amount of air you exhale in one forced breath?

c) Once you have made your prototype you need to calibrate it in some way. For example, you might measure the volume of air in one forced breath using a calibrated container inverted in a tank of water.

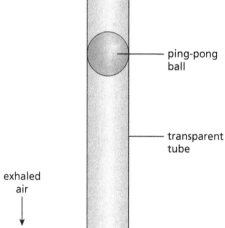

FIG 7.SIP.3 A simple design using a tube and a ping-pong ball

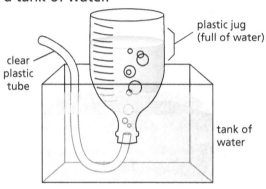

FIG 7.SIP.4 Measuring exhaled air

You can compare this to the reading on your device. You will need to do this with several people who exhale different volumes of air to get different values. This will allow you to mark a scale on your device.

d) Once you have calibrated your device it is time to test it by comparing your results with those of an actual spirometer. The local health centre may take measurements for the members of your team. You can compare those with the measurements obtained using your device. This will allow you to assess how accurate your device is.

e) As a result of testing two things need to be considered:

- How easy is my device to use?
- How accurate is my device?

You should use the answers to these questions to guide you in making improvements to your device.

f) Your final task is to write instructions on how to use your spirometer. The plan is for your device to be used in different countries where English may not be the first language. Bearing this in mind, keep the language as simple as possible and use lots of diagrams and/or photographs so that a user will be able to follow your instructions even if he/she cannot read all of the text.

Unit 8: Space science

We are learning how to:
• describe the solar system
• understand developments in space science

Space science ≫

Less than a century ago space flight was only a dream. Nowadays, rockets of different types leave for space on an almost weekly basis. Soon it may be possible for everyone to travel in space.

The term 'Solar System' is derived from the Sun. This star as the centre of the Solar System radiates light and heat onto the planets that are in orbit around it. The planets are held in position by the forces of gravitational attraction between the planets and the Sun.

FIG 8.1.1 The Solar System

The Solar System consists of the Sun, eight planets and the natural satellites or moons that orbit these planets. Starting nearest the Sun, the names of the planets are Mercury, Venus, Earth, Mars, Jupiter, Saturn, Uranus and Neptune.

FIG 8.1.2 The asteroid belt

In addition to the eight planets, there is a band of much smaller bodies of rock in orbit around the Sun between Mars and Jupiter. These are collectively called the asteroids, planetoids or minor planets. There are millions of asteroids which range in size from the largest Ceres, which has a diameter of 933 km, to some less than 1 km in diameter.

FIG 8.1.3 Stars and galaxies

Far beyond the edge of the Solar System there are stars, and collections of stars which we call galaxies. The bodies are so far away from the Earth that it is impractical to measure the distances in kilometres. Instead we measure distance in light years – the distance a ray of light can travel in one year.

Everything in space is contained within the Universe. Scientists think that the Universe is still expanding so it is not possible to say how big it is.

Fun fact

The Ancient Greeks believed the Sun rose and set each day because the sun god Helios drove his sun chariot across the sky.

The Solar System

We are learning how to:

- name and identify the components of the Solar System
- appreciate some of the properties of the planets in the Solar System.

Solar System ›››

The Sun is at the centre of the Solar System. There are eight planets and the asteroid belt in orbits moving around the Sun. There are also **natural satellites** or moons in orbits moving around some of the planets. Table 8.2.1 provides some data on the different planets.

Planet	Diameter at the equator (km)	Density (g/cm³)	Composition	Distance from the Sun (10⁶ km)	Time to orbit the Sun (years)
Mercury	4879	5.43	rock	58	0.3
Venus	12 103	5.24	rock	108	0.6
Earth	12 756	5.52	rock	150	1.0
Mars	6794	3.98	rock	228	1.9
Asteroid belt					
Jupiter	142 985	1.33	liquid/gas	778	11.9
Saturn	120 536	0.70	gas	1427	29.4
Uranus	51 118	1.30	gas	2871	83.8
Neptune	49 528	1.76	gas	4500	163.7

TABLE 8.2.1

It will be seen from the data that the planets can be conveniently divided into two groups – the inner four planets and the outer four planets.

Group	Relative size	Relative density	Composition	Relative time to orbit the Sun
Inner four planets	Small	High	Rock	Short
Outer four planets	Large	Low	Liquid/gas	Long

TABLE 8.2.2

Activity 8.2.1

Plotting a graph of distance from the Sun against time to orbit

Here is what you need:

- Graph paper
- Sharp pencil
- Ruler.

Here is what you should do:

1. Draw a vertical (*y*-axis) from 0 to 5000 and label this 'Distance from the Sun (10^6 km)'.
2. Draw a horizontal (*x*-axis) from 0 to 12 and label this 'Time taken for one orbit of the Sun (years)'.
3. Plot the data for the first five planets.
4. What pattern is there between the distance from the Sun and the time taken to orbit the Sun?
5. The minor plant Ceres is 414×10^6 km from the Sun. Use your graph to estimate how long it takes Ceres to complete one orbit of the Sun.

Natural satellites

In the same way that the planets are in orbit around the Sun, some planets have natural satellites or moons in orbit around them. These are also part of the Solar System.

The Earth has one natural satellite which we call the Moon. Some smaller planets have no moons while other larger planets have many. Although Mars is smaller than Earth, it has two moons, which are called Deimos and Phobos.

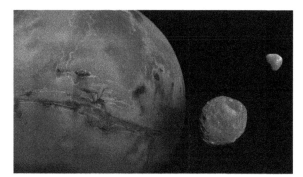

FIG 8.2.1 Mars has two moons

Check your understanding

1. Prior to 2006 the Solar System was deemed to have nine planets. The ninth 'planet' was called Pluto. It was downgraded to a dwarf planet. Here is some information about Pluto.

Diameter at the equator	2390 km
Density	1.86 g/cm³
Composition	Rock
Distance from the Sun	5906×10^6 km
Time taken to orbit the Sun	248 years

TABLE 8.2.3

Compare this information with the data given in Table 8.2.1 and state in what ways is Pluto similar to:

a) the inner four planets

b) the outer four planets.

Key terms

natural satellite another term for a moon

density the mass of 1 cm³ or 1 m³ of a substance

The inner four planets

We are learning how to:

- name and identify the inner four planets of the Solar System
- describe some of the properties of the inner four planets.

The inner four planets ⟩⟩⟩

In the previous lesson you learned that the inner four planets of the Solar System are smaller, denser and orbit the Sun more quickly than the outer four planets. The inner four planets are also composed of rock.

Although the four inner planets share these similarities, conditions on the planets are very different.

Mercury

Mercury is the smallest planet and it is nearest the Sun so you might expect conditions to be hot. The **surface temperature** at the equator can fall as low as −170 °C at night and can rise as high as 430 °C during the day. Mercury has no **atmosphere** and the surface is covered by craters resulting from the impact of **meteors**.

Venus

Venus is the planet that is closest in size to the Earth, but that is where the similarities end. You might expect the surface temperature of Venus to be less than that of Mercury because it is further from the Sun; however, it is much higher.

The reason for this is the composition of the planet's atmosphere, which consists of over 96% carbon dioxide. Carbon dioxide is a greenhouse gas. The gas traps heat emitted from the surface of the planet instead of allowing it to pass into space.

Earth

The Earth is sometimes called the 'blue planet' because of its appearance from space. It has an average surface temperature of 7 °C and an atmosphere which contains about one-fifth oxygen. In most of the areas of the Earth water exists as a liquid. The conditions on the surface of the Earth are ideal for life to exist.

Mars

Mars is sometimes called the 'red planet' because of its appearance at certain times of the year. Early astronomers noticed that the surface of the planet seemed to change

FIG 8.3.1 Mercury

FIG 8.3.2 Venus

FIG 8.3.3 Earth

FIG 8.3.4 Mars

colour over time. It was suggested that this corresponded to plants growing and dying off, much like what happens in the spring and the autumn on Earth. This observation was later explained by the presence of huge sand storms that sometimes spread over the surface of the planet.

The Mars Explorer project is currently gathering data on the surface of Mars to determine where there is or ever has been life on the planet.

FIG 8.3.5 Mars Explorer

Activity 8.3.1

Evidence for the runaway greenhouse effect on Venus

It is possible to evaluate the impact of the greenhouse effect on Venus from the pattern between the distance from the Sun and the average surface temperature for the inner four planets.

Planet	Mercury	Venus	Earth	Mars
Average surface temperature (°C)	167	457	7	–45
Distance from the Sun (10^6 km)	58	108	150	228

TABLE 8.3.1

Here is what you need:

- Graph paper   • Sharp pencil   • Ruler.

Here is what you should do:

1. Draw a vertical (y-axis) from –50 to 200 and label this 'Average surface temperature (°C)'.

2. Draw a horizontal (x-axis) from 0 to 250 and label this 'Distance from the Sun (10^6 km)'.

3. Plot the data for Mercury, Earth and Mars.

4. Draw the best curve you can through the three plots.

5. Estimate what the average surface temperature on Venus would be if it followed the same pattern as the other three planets.

6. Estimate by how much the average surface area of Venus has increased as a result of the greenhouse effect.

Check your understanding

1. Which of the inner four planets:

   a) is nearest the Sun?

   b) is closest in size to the Earth?

   c) has an average surface temperature that is much higher than would be expected from its distance from the Sun?

   d) appears blue when viewed from space?

2. State three conditions which make the surface of the Earth suitable for life.

Key terms

surface temperature the temperature on the surface of a planet

atmosphere layer of gases surrounding a planet

meteor lump of rock from space that enters the atmosphere of a planet and may collide with it

The outer four planets

We are learning how to:

- name and identify the outer four planets of the Solar System
- describe some of the properties of the outer four planets.

The outer four planets ⟫

In lesson 8.2 you learnt that the outer four planets of the Solar System are larger, less dense and orbit the Sun more slowly than the inner four planets. The outer four planets are composed of substances which are liquids and gases on Earth but because of the very low temperatures on these planets, some of the substances exist as liquids and solids.

FIG 8.4.1 Jupiter

Jupiter

Jupiter is the largest of all the planets. It has a large red spot which has intrigued observers for hundreds of years. This is believed to be caused by huge storms in the planet's atmosphere.

The average surface temperature of Jupiter is around −150 °C. It is called a **gas giant** planet because its atmosphere is composed mainly of hydrogen gas and helium gas, much like the Sun. What exists below the atmosphere is still a mystery to scientists but liquid hydrogen and helium are thought to be likely.

Saturn

Saturn is the second largest planet. It is surrounded by what appear to be several rings. Close examinations by space probes have shown that there are in fact many rings consisting of particles of different sizes.

FIG 8.4.2 Saturn

The atmosphere of Saturn is similar to that of Jupiter in consisting of about 75% hydrogen and 25% helium. Beneath the atmosphere are substances like ammonia, water and methane, all in solid form.

Uranus

Uranus is similar in size to Neptune and significantly smaller than Jupiter and Saturn. Unusually the **axis of rotation** of this planet points towards the Sun rather than being perpendicular to it. This gives rise to some doubt over which end of the axis should be regarded as the north, and which end the south. It has rings similar to Saturn but they are much less prominent.

The atmosphere of Uranus consists of hydrogen and helium together with around 2% **methane**. The presence of methane in the atmosphere is responsible for its blue colour. The core of the planet is thought to be composed of substances like water, methane and **ammonia** in a liquid/solid state.

FIG 8.4.3 Uranus

Neptune

Neptune is the outermost planet of the Solar System. It is similar in size and structure to Uranus and is sometimes described as an ice giant. It is so far away from the Sun that the average surface temperature is around −200 °C.

Less is known about the outer planets because of the greater distance from the Earth. Pioneer 10 was the first space mission to the outer planets and it was launched in 1973.

The outer four planets have considerably more moons than the inner four planets. Table 8.4.1 shows the number of known moons as of 2017.

It is likely that these numbers will increase as improving rocket technology allows a more detailed examination of these planets.

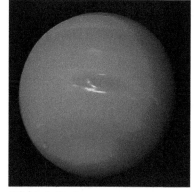

FIG 8.4.4 Neptune

Activity 8.4.1

Exploring the outer four planets

There have been a number of projects to explore the outermost planets including the Galileo probe (launched in 1989), the Cassini spacecraft (launched in 2004) and the Juno spacecraft (launched in 2011).

Here is what you should do:

1. Choose one of the above three projects to explore the outer planets.

2. Carry out research on the project using whatever resource materials are available. You should record details such as:

   a) the purpose of the project

   b) which planet(s) were observed

   c) what sort of information was gathered.

3. Prepare a short summary of what you find that you can share with other students in the class.

Planet	Number of natural satellites or moons
Jupiter	67
Saturn	62
Uranus	27
Neptune	14

TABLE 8.4.1

Key terms

gas giant large planet consisting mostly of gas

axis of rotation axis between the North and South Poles around which a planet rotates

methane a gas under normal conditions on Earth which has the formula CH_4

ammonia a gas under normal conditions on Earth which has the formula NH_3

Check your understanding

1. Which of the outer four planets:

   a) is surrounded by prominent sets of rings?

   b) has a large red spot?

   c) appears blue due to the presence of methane in the atmosphere?

   d) is the largest planet in the Solar System?

Gravity

We are learning how to:

• define gravity as a force that pulls objects towards the centre of the Earth
• explain the role of gravity in keeping bodies in orbit around other bodies in space.

Gravity ⟫

Gravity is the force that holds objects on the Earth and pulls objects towards the ground. If you drop an object it will fall to the ground.

Although Sir Isaac Newton was not the first to propose the idea of gravity, he reasoned that any object which had mass would exert a gravitational pull. If two bodies each exerted a gravitational pull they would attract each other. Newton proposed mathematical relationships to explain the forces of attraction between objects.

The size of the force of attraction between two objects depends on:

FIG 8.5.1 Newton contributed greatly to our understanding of gravity

- the masses of the objects – the greater the masses the greater the force of attraction

- the distance between the objects – the closer the objects the greater the force of attraction.

A force of attraction exists between you and the person sitting next to you. Even though you are sitting quite close to each other your masses are very small so the force of attraction between you is very small.

When the masses of the objects are very large, the force of attraction between them is also very large.

The masses of the Sun and the Earth are very large. Even though they are 150 million kilometres apart the force of attraction between the Sun and the Earth is around 3.5×10^{23} N. This is enough to keep the Earth in orbit around the Sun and prevent it from flying off into space.

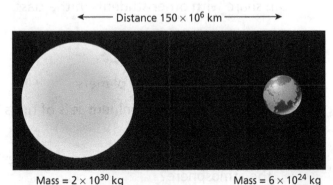

← Distance 150×10^6 km →

Mass = 2×10^{30} kg   Mass = 6×10^{24} kg

FIG 8.5.2 There is a large force of attraction between the Sun and the Earth

Activity 8.5.1

Modelling the motion of a planet around the Sun

Here is what you need:

- Small ball or weight
- Plastic tube wide enough for the cord to pass through
- Nylon cord
- Washer.

Here is what you should do:

1. Tie one end of the nylon cord to the ball.

2. Pass the cord through the plastic tube and tie the other end to the washer.

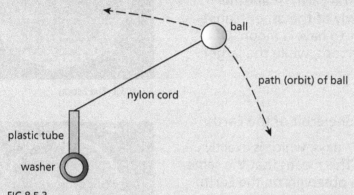

FIG 8.5.3

3. Hold the washer and plastic tube and swing the ball around so it is moving around the top of the plastic tube.

4. Experiment by moving the washer so that the distance between the ball and the top of the plastic tube becomes less.

5. Does altering the distance have any effect on the speed with which the ball moves around?

Check your understanding

1. What force prevents the Earth from leaving its orbit around the Sun?

2. Fig 8.5.4 gives information about the Earth and the Moon.

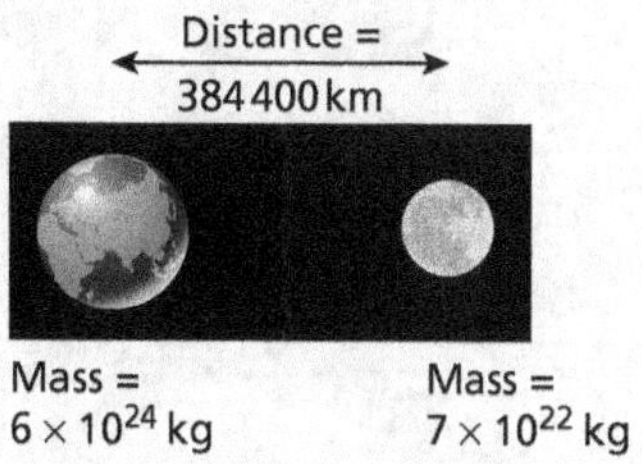

Distance = 384 400 km

Mass = 6×10^{24} kg

Mass = 7×10^{22} kg

FIG 8.5.4

Compared to the magnitude of the force of attraction between the Sun and the Earth, why might you think that the force of attraction between the Earth and the Moon is:

a) larger?　　b) smaller?

Key terms

gravity force or attraction exerted by anything which has mass

The Moon

We are learning how to:

• describe the motion of the Moon about the Earth.

Natural satellites ›››

In the same way as the Earth moves in an orbit around the Sun, so the Moon moves in orbit around the Earth.

The Moon can be described as a natural **satellite** of the Earth. A satellite is a body orbiting the Earth or another planet. You will recall from your study of the other planets that the Earth is not the only planet to have a moon. Of the smaller planets, Mars has two moons, while the larger planets each have many moons.

FIG 8.6.1 The Moon

Orbit of the Moon

It takes the Moon 27 days to make one orbit of the Earth.

The Moon rotates on its axis every 27 days, which is exactly the same time as it orbits the Earth. This means that the same area of the Moon is always visible to observers on the Earth.

Phases of the Moon

The Moon appears to change shape during its orbit around the Earth due to the way in which it reflects light from the Sun. These different shapes are called the **phases of the Moon**.

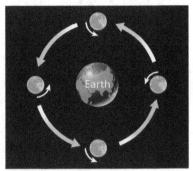

FIG 8.6.2 The Moon rotates as it moves around the Earth

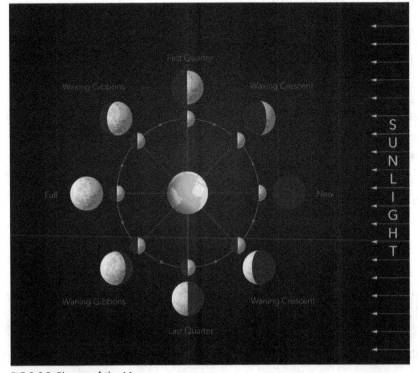

FIG 8.6.3 Phases of the Moon

The Moon is full when it is on the opposite side of the Earth to the Sun. In this position it acts like a huge mirror reflecting sunlight onto the Earth.

Moon landings

The first person to step onto the Moon was Neil Armstrong on 20 July 1969. He was one of three astronauts on the Apollo 11 mission. The other crew members were Edwin 'Buzz' Aldrin, who also walked on the Moon, and Michael Collins, who remained in orbit around the Moon.

Subsequent to the Apollo 11 mission, the Moon was visited again by the manned Apollo missions 12, 14, 15, 16 and 17.

FIG 8.6.4 Apollo 11 was the first spaceship to land on the Moon

Activity 8.6.1

Motion of the Moon around the Earth

You should work in pairs to carry out this activity.

Here is what you need:

- Flashlight (to represent the Sun)
- Small ball (to represent the Moon)
- Large ball (to represent the Earth).

Here is what you should do:

1. Darken the room to carry out the activity.
2. Place the flashlight on the desk to represent the Sun.
3. One person should hold the Earth in place in front of the Sun.
4. The other person should move the Moon into different positions around the Earth to observe the different phases.
5. Is the Moon visible from the Earth when the Sun, Earth and Moon are exactly in line?

Fun fact

In the past the 'dark side of the Moon', i.e. the part that is not visible from the Earth, was the subject of much speculation by science fiction writers. However, it was photographed by the Soviet space probe Luna 3 in 1959 and, not surprisingly, its appearance was very similar to the visible side.

Check your understanding

1. Why does the same side of the Moon always face the Earth?
2. Where is the Moon relative to the positions of the Earth and the Sun during a 'new moon'?
3. Who was the first person to set foot on the surface of the Moon?

Key terms

satellite celestial body in orbit around a planet

phases of the Moon changes in the apparent shape of the Moon at different times during its orbit

Light and shadow

We are learning how to:

- determine whether a light source is luminous or non-luminous
- determine whether a material is transparent, translucent or opaque
- describe the formation of shadows.

Light sources

Light sources can be divided into two groups:

- **Luminous light sources** are those which actually produce light. The Sun and the other stars we can see in the night sky are luminous light sources.

- **Non-luminous light sources** are those which do not produce light themselves, but simply reflect the light produced by luminous sources. The Moon and the planets are non-luminous light sources. We can only see them because they reflect light from the Sun.

Although stars are luminous and planets are non-luminous they look similar in the night sky.

Transparent, translucent and opaque

Materials react with light in different ways.

We can put materials into groups according to how much light they absorb and how much light passes through them.

A **transparent** material lets most of the light pass through it.

A **translucent** material absorbs some light and also lets some light pass through it.

An **opaque** material absorbs all the light and lets none pass through it.

Shadows

Light travels in straight lines. It cannot pass through or bend around an opaque object.

Such an object casts a **shadow** on the side opposite from where it is illuminated. A shadow is a shape formed when the path of light is blocked by an opaque object.

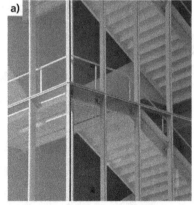

FIG 8.7.1 **a)** Glass is a transparent material

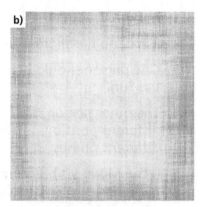

FIG 8.7.1 **b)** Cotton is a translucent material

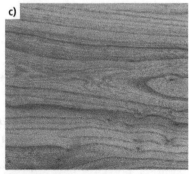

FIG 8.7.1 **c)** Wood is an opaque material

FIG 8.7.2 Shadows form because light travels in straight lines (if light was able to bend around an object, the object would not cast a shadow)

Activity 8.7.1

Investigating shadows

Here is what you will need:

- Flashlight with lens covered by a card with a hole in it
- Cardboard star shape
- Screen or area of lightly coloured wall
- Modelling clay
- Ruler.

Here is what you should do:

1. Place a flashlight on some books a metre or so in front of a screen.
2. Fix a star shape 10 cm from the screen using modelling clay.
3. Measure the height of the star.
4. Repeat steps 2 and 3 four times but moving the star away from the screen by an extra 10 cm each time.
5. Record the distance from the screen to the object and the height of the shadow each time in a table.

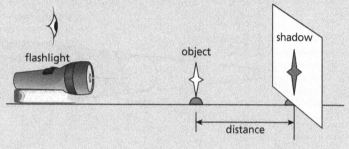

FIG 8.7.3

Check your understanding

1. Fig 8.7.4 shows an opaque object between a light source and a screen.

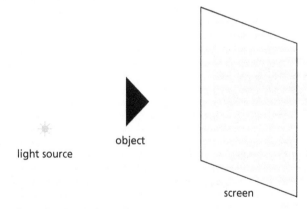

FIG 8.7.4

a) Copy Fig 8.7.4 and draw straight lines from the light source through the corners of the object to show the shadow that is formed on the screen.

b) State one similarity and one difference between the object and its shadow.

Key terms

luminous light source one which generates light

non-luminous light source one which reflects light produced by a luminous source

transparent material that allows most light to pass though it

translucent material that allows some light to pass through it

opaque material that does not allow any light to pass through it

shadow area behind an opaque object which does not receive light

Partial and full shadow

We are learning how to:

- explain why areas of partial shadow and areas of full shadow may form.

Partial and full shadow ⟩⟩⟩

A **point light source** creates a single area of shadow behind the object.

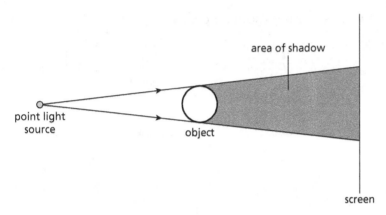

FIG 8.8.1 Shadow created by a point light source

When a large or **extended light source** is used, two different areas of shadow are created. This is possible because each side of the extended light source sends rays of light both above and below the object.

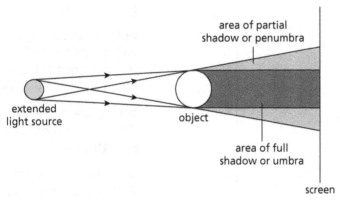

FIG 8.8.2 Shadow created by an extended light source

An area of full shadow, called the **umbra**, is created immediately behind the object. Surrounding the umbra there is an area of partial shadow called the **penumbra**.

Activity 8.8.1

Investigating the shadow formed by an extended light source

Here is what you need:

- Flashlight
- Cardboard circle shape
- Screen or area of lightly coloured wall
- Ruler
- Modelling clay.

Here is what you should do:

1. Place a flashlight on some books a metre or so in front of a screen.
2. Hold a circular shape about 10 cm from the screen.
3. Move the object back and forth until you can see an area of full shadow and an area of partial shadow surrounding it. Fix the object in place using modelling clay.

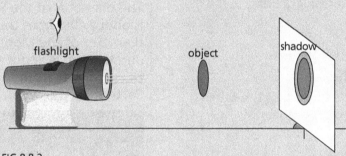

flashlight   object   shadow

FIG 8.8.3

4. Measure the diameter of the area of full shadow.
5. Measure the width of the band of partial shadow.
6. Make a labelled diagram of the shadow cast by the object.

> **Fun fact**
>
> Objects often cast areas of full and partial shadow when illuminated by everyday light sources such as lamps.
>
> However, we are often not aware of them because the area of partial shadow appears as no more than a fuzzy border around the area of full shadow.

Check your understanding

1. A student is sitting under a small lamp. He is finding it difficult to read a book. Answer the following questions in terms of the shadows formed.

   a) Explain why, in terms of light and the formation of shadows, the student is finding it difficult to read.

   b) What difference would it make if the small lamp was replaced by a long fluorescent strip light?

FIG 8.8.4

Key terms

point light source source of light that creates a single area of shadow behind the object

extended light source source of light that creates two different areas of shadow

umbra area of full shadow

penumbra area of partial shadow

Solar eclipse

We are learning how to:
- explain why solar eclipses take place.

Solar eclipse >>>

The Sun acts as an extended source of light. The Earth orbits the Sun and the Moon orbits the Earth. Sometimes the Sun, the Earth and the Moon line up in a straight line.

A **solar eclipse** takes place when the Moon passes between the Sun and the Earth. The eclipse is the result of the shadow cast by the Moon passing over the Earth.

Activity 8.9.1

Modelling a solar eclipse

You should carry out this activity in a small group.
Here is what you need:

- Flashlight
- Large ball
- Small ball
- Marker.

The flashlight represents the Sun, the large ball the Earth and the small ball the Moon. Different members of the group will hold each one.

Here is what you should do:

1. Draw an 'X' on the large ball to represent the position of Jamaica.

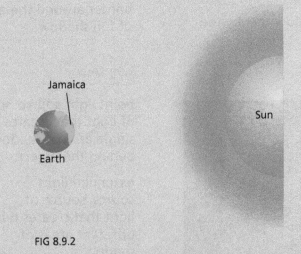

FIG 8.9.2

2. Place the Earth about 2 m from the Sun.
3. Bear in mind that the Moon is much nearer to the Earth than it is to the Sun.

Fun fact

The diameter of the Sun is about 400 times the diameter of the Moon and the Sun is about 400 times further from the Earth than the Moon.

FIG 8.9.1 The Sun and Moon appear to be similar in size when viewed from the Earth

This is why the Sun and the Moon look to be very similar in size when we see them in the sky.

Key term

solar eclipse when the Moon passes between the Sun and the Earth, the Moon casts a shadow as it passes across the Earth

4. Place the Moon in different positions between the Earth and the Sun until you find a place where the Moon casts a shadow on Jamaica.

5. Redraw Fig 8.9.2 showing the position of the Moon.

During a solar eclipse the Moon passes in front of the Sun.

The Sun is so large compared to the Moon that the Moon casts a region of full shadow, called the umbra, and a region of partial shadow called the penumbra on the Earth.

What people see on Earth during a solar eclipse depends on where they are.

- People in the umbra will see a total eclipse.

- People in the penumbra will see a partial eclipse.

- People outside the penumbra will not see any eclipse.

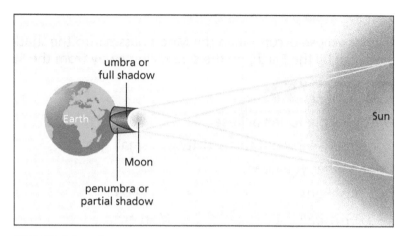

FIG 8.9.3 Solar eclipse

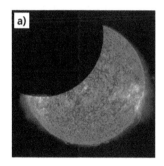 

FIG 8.9.4 **a)** Partial eclipse of the Sun **b)** Total eclipse of the Sun

Check your understanding

1. Fig 8.9.5 shows the Sun, the Moon and the Earth.

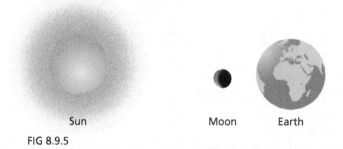

Sun   Moon   Earth

FIG 8.9.5

a) Which eclipse takes place when the three bodies are in this position?

b) Copy and complete Fig 8.9.5 by drawing lines to show the shadow cast by the Moon on the Earth.

c) Label the umbra and the penumbra.

Lunar eclipse

We are learning how to:

- explain why lunar eclipses take place.

Lunar eclipse >>

A **lunar eclipse** occurs when the Moon passes into the shadow produced by the Earth, on the side facing away from the Sun.

Activity 8.10.1

Modelling a lunar eclipse

You should carry out this activity in a small group.

Here is what you need:

- Flashlight
- Large ball
- Small ball
- Marker.

The flashlight represents the Sun, the large ball the Earth and the small ball the Moon. Different members of the group will hold each one.

Here is what you should do:

1. Draw an 'X' on the large ball to represent the position of Jamaica.

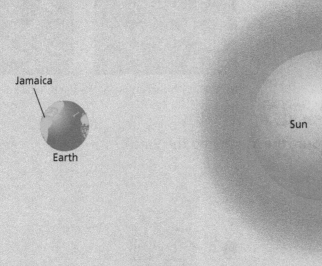

FIG 8.10.1

2. Place the Earth about 2 m from the Sun.

3. Bear in mind that the Moon is much nearer to the Earth than it is to the Sun.

4. Place the Moon in different positions around the Earth until you find a place where the Moon is in the shadow of the Earth and could not be seen from Jamaica.

5. Redraw Fig 8.10.1 showing the position of the Moon.

During a lunar eclipse the Moon passes into the Earth's shadow. As the Moon does not produce its own light, it looks even darker in the Earth's shadow during a lunar eclipse.

The Moon may pass through a region of partial or a region of full shadow.

- When the Moon passes through the Earth's penumbra we see a partial eclipse.

- When the Moon passes through the Earth's umbra we see a total eclipse.

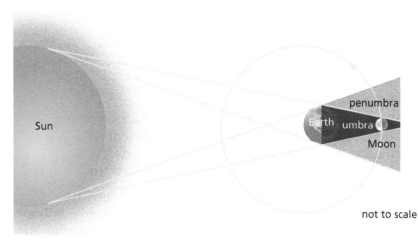

not to scale

FIG 8.10.2 Lunar eclipse

FIG 8.10.3 **a)** Partial eclipse of the Moon **b)** Total eclipse of the Moon

In fact, the Moon never becomes totally invisible. This is because some light is refracted towards it by the Earth's atmosphere. During a total lunar eclipse the Moon often looks red as a result of this refracted light.

Check your understanding

1. **a)** Draw a diagram to show the positions of the Sun, Earth and Moon during a lunar eclipse.

   b) Explain why the appearance of the Moon, as seen from Earth, becomes much darker during a lunar eclipse.

Fun fact

Islamic astronomers living over a thousand years ago knew enough about the movement of the Earth and the Moon to be able to predict accurately when eclipses would happen.

Key term

lunar eclipse when the Moon passes into the shadow produced by the Earth

Other objects in the Solar System

We are learning how to:

- explore within the Solar System
- describe meteors and comets.

Meteors

Meteoroids are pieces of debris that are found within the Solar System in space between planets. Meteoroids can be as small as a grain of sand or as large as a boulder. Pieces of debris which are larger than this are called **asteroids**.

As the Earth orbits the Sun it continually collides with meteoroids in its path. The meteoroids only become visible when they pass into the Earth's atmosphere. At this time the tracks of light they produce are called **meteors**.

Meteors are visible because meteoroids heat up as they pass through the air. Most meteors only glow for a second or two before they vaporise and disappear. Very rarely a meteoroid may be large enough to survive passing through the atmosphere and it will strike the surface of the Earth. It is then called a **meteorite**.

FIG 8.11.1 Meteors produce lots of light

Comets

Comets are small objects that orbit around the Sun but their orbit is very different from that of a planet. A comet has a much more elliptical orbit. Part of the orbit of a comet is outside the Solar System.

> ### Fun fact
>
> Halley's Comet appears as a very bright light in the sky as it passes close to the Earth; however, it can only be seen every 76 years and will not be visible again until 2061.

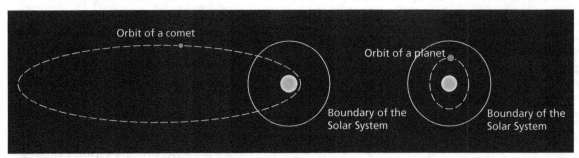
FIG 8.11.2 Part of the orbit of a comet is outside the Solar System

The centre or nucleus of a comet consists of a mixture of ice, dust and rock particles and can be anything from a few hundred metres to tens of kilometres in diameter.

As a comet approaches the Sun, volatile materials vaporise and stream out of the nucleus carrying dust particles with them. Radiation from the Sun causes a 'tail' to form. This tail always points away from the Sun.

Most comets are too faint to be seen without the aid of a telescope but very occasionally, a comet like Halley's Comet appears, which is visible to the naked eye. The orbital time of a comet may be anything from a few years to hundreds of thousands of years. At the start of 2011 there were 4185 reported comets.

Activity 8.11.1

Drawing the orbit of a comet

The orbit of a comet is elliptical and not circular so you cannot draw it with compasses. A different method is needed.

Here is what you need:

- Plain paper
- Drawing board
- String
- Thumb tack × 2.

Here is what you should do:

1. Place your paper on a drawing board so that you can push drawing pins into it.
2. Draw a Sun at one end of your piece of paper.
3. Place the drawing pins in a line with the Sun.
4. Tie the string in a loop long enough to reach from the left-most thumb tack to the far side of the Sun.

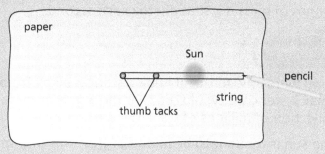

5. Place the end of your pencil at the right end of the loop and draw a line, keeping the string taut all the time. If it runs off the paper you will need to reposition your thumb tacks.

FIG 8.11.3

6. What shape have you drawn?
7. Mark the position of a comet on the orbit you have drawn.
8. If you have time, experiment by moving the left-most thumb tack closer to and further from the other thumb tack and see what effect this has on the shape of the orbit.

Check your understanding

1. State two ways in which the orbit of a comet differs from that of a planet.
2. There are many meteoroids in space but very few meteorites land on the Earth. Explain why.

Key term

meteoroid small particles of debris within the Solar System

asteroid large meteoroid

meteor meteoroid once it enters the Earth's atmosphere

meteorite meteor which lands on the Earth

comet body that orbits the Sun but spends some time outside the Solar System

Objects beyond the Solar System

We are learning how to:

- explore beyond the Solar System
- describe stars and galaxies.

Stars ⟩⟩

The Sun is a star at the centre of the Solar System. It has existed for billions of years and will continue to shine for billions of years to come. Although the Sun is massive compared with the planets, it is quite modest in size compared with many other stars.

All stars are luminous light sources. They produce heat and light as a result of a process called nuclear fusion. Under the extreme conditions on a star, small atoms combine to form larger atoms with the loss of a small amount of mass. This mass is converted into energy. On our Sun, over 4 000 000 tonnes of mass is lost as energy each second.

FIG 8.12.1 The Sun is one star in a spiral galaxy

Galaxies ⟩⟩

If we were able to travel huge distances away from the Solar System, we would see that the Sun is just one star in a collection of millions of stars called a **galaxy**.

The Sun is thought to be in one of the arms of a **spiral galaxy**. When we observe the **Milky Way** in the night sky we are looking across this galaxy and what we are really seeing is the millions of stars that make up the galaxy.

There are galaxies which are other shapes.

FIG 8.12.2 **a)** Spiral galaxy   **b)** Elliptical galaxy

In an **elliptical galaxy** the stars form an elliptical shape. In an **irregular galaxy** there is no obvious pattern to the distribution of stars.

> **Fun fact**
>
> All stars appear to be white but on closer examination they radiate a range of colours according to how hot they are. The coolest stars radiate red light while the hottest stars radiate white/blue light.

If we could continue to travel even further away from our galaxy we would see that there are millions of other galaxies that are different in shape and size. In the same way that a galaxy is composed of stars, so the galaxies themselves may form a cluster. **Galaxy clusters** contain billions of stars.

A **super cluster** is a large group of smaller galaxy clusters; they are the largest known structures in the Universe. A super cluster contains many millions of stars.

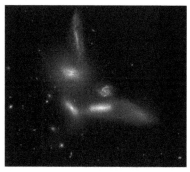

FIG 8.12.3 A galaxy cluster may contain many galaxies

Activity 8.12.1

Constellations

You will not need any apparatus for this activity.

Constellations are groups of stars which were thought by ancient observers to resemble animals or characters from mythology. The pattern of stars in Fig 8.12.4 was named after the hunter Orion.

Use whatever resources are available to investigate what other constellations are visible in the night sky over Jamaica. If you have internet access, a good place to start might be the website of the Astronomical Association of Jamaica.

Key term

galaxy collection of stars

spiral galaxy millions of stars arranged in a spiral shape

Milky Way the spiral galaxy of which the Sun is a star

elliptical galaxy millions of stars arranged in an elliptical shape

irregular galaxy millions of stars not arranged in any obvious shape

galaxy cluster collection of galaxies

super cluster group of galaxy clusters

Check your understanding

1. Arrange the following in terms of size starting with the smallest:

   cluster galaxy star super cluster Universe

Distance in space

We are learning how to:

- appreciate the relationship between the Universe, a galaxy and a star
- express distances in space.

The Universe

According to the **Big Bang Theory**, the **Universe** was created 13 billion years ago as a result of a massive explosion at a single point.

Matter was thrown in all directions and according to scientists objects in space are still moving away from each other. This means that the Universe is still expanding.

The Universe consists of all matter and space. Within the Universe there are many galaxies, and within a galaxy there are many stars.

FIG 8.13.1 The birth of the Universe

Distances in space

With the exception of the Sun, the nearest star to Earth is a long way away. This star is called Proxima Centauri and it is about 4.3×10^{13} or 43 000 000 000 000 kilometres away. To put this into perspective, if you were sat in a classroom in Morant Bay and the person next to you was the Sun, Proxima Centauri would be in Negril.

In order to avoid using large numbers when describing the positions of objects in space, scientists use a different unit for measuring distance.

Light travels at a constant speed of 300 000 000 m/s in a vacuum. In one year a beam of light will therefore travel:

300 000 000 × 60 × 60 × 24 × 365 = $9.46 \times 10^{12} \approx 10^{13}$ kilometres

This value is similar in magnitude to the nearest stars to Earth, therefore distance in space can be expressed more sensibly in light years. The five stars nearest Earth are shown in the table below.

Star	Distance from Earth in light years
Proxima Centauri	4.24
α Centauri A	4.37
α Centauri B	
Barnard's Star	5.96
Wolf 359	7.78

TABLE 8.13.1

The furthest stars from our Solar System are thought to be billions of light years away but nobody really knows because the exact size of the Universe is unknown. At present it is thought to be at least 10 billion light years in diameter.

Activity 8.13.1

How long does sunlight take to reach the planets?

Table 8.13.2 shows the distance from the Sun to all of the planets in the Solar System.

Planet	Distance from the Sun (10^6 km)	Time taken for light to travel from the Sun to the planet
Mercury	58	
Venus	108	
Earth	150	$\dfrac{150\ 000\ 000\ 000}{300\ 000\ 000} = 500$ s $= 8$ min 20 s
Mars	228	
Jupiter	778	
Saturn	1427	
Uranus	2871	
Neptune	4500	

TABLE 8.13.2

Here is what you should do:

1. Copy and complete the table to show how long it takes for light to travel to each of the planets in the Solar System (the Earth has already been done for you).

2. Express each time in appropriate units.

> **Fun fact**
>
> Distance in space is also measured in another even bigger unit called the parsec. There are 3.26 light years in 1 parsec.

Check your understanding

1. **a)** What is the nearest star to the Earth excluding the Sun?

   b) How far away is it in:

   i) kilometres?   **ii)** light years?

   c) A spaceship is able to travel through space at about 58 000 kilometres per hour. Travelling in this spaceship, how long would it take to reach this star?

> **Key terms**
>
> **Big Bang Theory** the Universe was created by a massive explosion
>
> **Universe** all existing matter and space

Review of Space science

- The Solar System consists of the Sun, the planets, the moons in orbit around the planets and the asteroid belt.
- The Sun is at the centre of the Solar System.
- Moons are sometimes called natural satellites.
- The inner four planets are Mercury, Venus, Earth and Mars.
- The inner four planets are relatively small, have a relatively high density, are made of rock and take a short time to orbit the Sun.
- The outer four planets are Jupiter, Saturn, Uranus and Neptune.
- The outer four planets are relatively large, have a relatively low density, are made of liquids and gases and take a long time to orbit the Sun.
- Gravity is a force which attracts objects towards each other.
- The planets are held in orbit around the Sun due to gravitational attraction, and moons are held in orbit around planets.
- The size of the gravitational force between two objects depends on their masses and the distance between them.
- The Moon is a natural satellite in orbit around the Earth.
- The Moon makes an orbit of the Earth every 27 days.
- The Moon rotates on its axis at the same speed as it orbits the Earth so the same area of the Moon is always visible from Earth.
- The Moon appears to change shape during its orbit due to different amounts being illuminated by the Sun. These are called the phases of the Moon.
- The first Moon landing was made by the astronaut Neil Armstrong on 20 July 1969.
- Luminous sources produce light; the Sun is a luminous light source.
- Non-luminous sources do not produce light but reflect light from luminous sources; the Moon is a non-luminous light source.
- Transparent materials allow most light to pass through them.
- Translucent materials allow some light to pass through them.
- Opaque materials allow no light to pass through them.
- A shadow is a region where light does not reach.
- An umbra is a region of total shadow and a penumbra is a region of partial shadow.
- A solar eclipse occurs when the Moon passes between the Sun and the Earth.
- A lunar eclipse occurs when the Earth passes between the Sun and the Moon.
- Meteors are pieces of rock that enter the Earth's atmosphere and burn up with a bright light.
- Comets are small bodies in orbit around the Sun and which pass out of the Solar System for part of their orbit.
- A galaxy is a large group of stars.

- The Sun is a star in a spiral galaxy which is called the Milky Way.
- A cluster is a group of galaxies.
- Everything in space is contained within the Universe.
- Distance in space is measured in light years.
- A light year is equal to a distance of approximately 10^{13} kilometres.
- The nearest star to the Earth after the Sun is Proxima Centauri, which is 4.24 light years away.

Review questions on Space science

1. **a)** Name the four inner planets in order starting with the one nearest the Sun.

   b) Name the four outer planets in order starting with the one furthest away from the Sun.

2. Which of the planets in the Solar system:

   a) has exactly two moons? **b)** has a giant red spot?

   c) has the greatest diameter? **d)** is closest in size to the Earth?

   e) is similar in size to Neptune? **f)** orbits the Sun quickest?

   g) passes closest to the Earth?

3. **a)** What is gravity?

   b) Explain how gravity keeps the Moon in orbit around the Earth.

4. **a)** Give one example each in space of:

   i) a luminous light source **ii)** a non-luminous light source.

   b) Explain the terms transparent, translucent and opaque, when applied to materials, giving one example each.

   c) Explain why an opaque shape creates a shadow but a transparent one does not.

5. **a)** What is the difference between an umbra and a penumbra?

   b) Why do some light sources create only an umbra while others also create a penumbra?

6. Draw a labelled diagram to show the positions of the Moon, Earth and Sun during a solar eclipse.

7. **a)** Between which two planets is the asteroid belt?

   b) Explain the terms meteoroid, meteor and meteorite.

8. **a)** What is a:

   i) galaxy? **ii)** galaxy cluster?

   b) Why are distances in space often given in light years?

   c) Calculate how many kilometres light travels in 1 year of 365 days.

Space science

There is an open day planned at your school during which parents and educationalists will be invited to come and see the sort of work that students carry out.

Your Principal has decided she would like each department to create a special display of some kind that the visitors will find interesting.

As you have just been studying the Solar System, the head of the science department has asked you to create a model of the Solar System in the school compound. In addition to this you will take groups of visitors on a walk from the Sun past the planets. As you pass each planet you will provide some information about it.

FIG 8.SIP.1 An example of a scale model of planets hanging in the hallway

1. You are going to work in groups of 3 or 4 to create a model of the Solar System in the school compound. The tasks are:

- To review the Solar System and in particular:
  - the order of the planets starting with the one that is nearest to the Sun
  - the relative size of each planet
  - the relative distance from the Sun to each planet
  - the important features of each planet.
- To consider a suitable scale for the distances from the Sun to the planets in your model.
- To provisionally mark out the positions of the planets in order to ensure that your Solar System will fit into the available space.
- To consider a suitable scale for the sizes of the planets in your model.
- To calculate the diameters of all of the planets using your scale to ensure they are a sensible size.
- To construct your planets and locate them in the space available.
- To create a short commentary, which will be used to inform guests about the planets as you walk from the Sun to the outermost planet, Neptune.

a) Look back through the unit and make sure you are familiar with the important features of the eight planets, and in particular their order starting at the Sun.

b) How are you going to decide on a suitable scale for the distances from the Sun to each of the planets? This is going to be determined by how much room is available in the school compound.

Suppose that you have a school field that is 100 m long and the Sun is at one end. The outermost planet is Neptune, which is 4500×10^6 km from the Sun. If we decide that Neptune is to be 90 m from the Sun then:

4500×10^6 km is equivalent to 90 m therefore $\dfrac{4500 \times 10^6}{90} = 50 \times 10^6$ is equivalent to 1 m

You will need to draw up a similar table according to the space you have available.

c) The next step is to provisionally mark out the positions of the planets to make sure your Solar System is going to fit the space available.

You need to mark out the positions of the planets. Perhaps you could make name flags using sticks and white card.

d) How are you going to decide on a suitable scale for the sizes of the planets? Rather than using the absolute values for the diameters, you might find it easier to determine their sizes relative to the diameter of the Earth.

You might be able to represent some of the planets by objects such as balls but you will need to be careful with your choice. For example, the diameter of Jupiter is 11 times that of Earth. If you decide to represent the Earth by a football 30 cm in diameter then you will need a sphere of diameter 11 × 30 = 3.30 m to represent Jupiter.

Similarly, if you decide to represent Jupiter by a football 30 cm in diameter then, on the same scale, the diameter of mercury will only be $\frac{30}{11} \times \frac{1}{3}$ = 0.9 cm, which will be too small to see clearly on your model.

You will probably be able to find suitably sized spheres for some planets but will need to make spheres for others. Small planets could be made from modelling clay while large planets could be made from chicken wire and paper mache.

e) Once your planets are built you need to decide the best way to display them. You might decide simply to place them on the ground but the smaller ones might be difficult to see.

You could place your planets on sticks.

f) Once you have completed your planets. Recheck your distances to make sure they are accurate and place your planets in their positions.

g) Your final task is to write a commentary.

You could take a picture of your Solar System from a first-floor window and use it to illustrate a handout of your commentary that you could give to your guests for them to take away.

Planet	Actual distance (10^6 km)	Distance in the model
Mercury	58	$\frac{58 \times 10^6}{50 \times 10^6}$ = 1.16
Venus	108	$\frac{108 \times 10^6}{50 \times 10^6}$ = 2.16
Earth	150	$\frac{150 \times 10^6}{50 \times 10^6}$ = 3.00
Mars	228	$\frac{228 \times 10^6}{50 \times 10^6}$ = 4.56
Jupiter	778	$\frac{778 \times 10^6}{50 \times 10^6}$ = 15.56
Saturn	1427	$\frac{1427 \times 10^6}{50 \times 10^6}$ = 28.54
Uranus	2871	$\frac{2871 \times 10^6}{50 \times 10^6}$ = 57.42
Neptune	4500	$\frac{4500 \times 10^6}{50 \times 10^6}$ = 90.00

TABLE 8.SIP.1

Planet	Actual diameter (km)	Approximate diameter compared with the Earth
Mercury	4900	$\frac{1}{3}$
Venus	12 100	1
Earth	12 800	1
Mars	6 800	$\frac{1}{2}$
Jupiter	143 000	11
Saturn	120 500	9
Uranus	51 100	4
Neptune	49 500	4

TABLE 8.SIP.2

FIG 8.SIP.2 Chicken wire frame ready for paper mache to be pasted on

Unit 9: Water and the Earth's atmosphere

We are learning how to:
- describe the surface of the Earth and its environment
- be aware of our water supplies.

Water and the Earth's atmosphere »

The Blue Planet

When viewed from space the Earth appears to be blue. This is because around 70% of its surface is covered in water. Water is perhaps the most essential substance on Earth. Every living thing, plants and animals, rely on water for their survival. Without water life would not have evolved on our planet.

Scientists are able to survey other planets and their surrounding moons using spaceships equipped with sophisticated detectors. They are interested in finding out if there is any water present now, or there has been in the past, because this might indicate the presence of life.

Over 99% of the water on Earth is in the seas and oceans. Sea water is called salt water because it contains dissolved salts like sodium chloride (table salt). It provides a habitat for many marine organisms but it cannot be consumed or absorbed by animals and plants living on land.

FIG 9.1.1 The Blue Planet

Fresh water

Each adult requires around 3 dm³ of fresh water each day to remain healthy. They get this through drinks and in their food. In addition fresh water has many other important uses such as washing, cleaning, sanitation and irrigation. Supplies of fresh water would have been exhausted many millions of years ago were it not for the water cycle.

FIG 9.1.2 Rain replenishes supplies of fresh water

The water cycle is a sequence of natural processes that replenish supplies of fresh water. We receive this as rain. The amount of rain you receive depends on where you live. In Kingston you can expect it to rain, on average, 93 days in the year, which will give you a total of about 810 millimetres.

Rain collects in streams, rivers and lakes. In some places it may be sufficiently pure to drink but more often, the water requires some treatment to ensure it does not contain harmful microorganisms.

Water treatment is a costly process. The demand for water can be reduced by finding ways of conserving water.

The atmosphere

The Earth is surrounded by a layer of gases we call the atmosphere. This extends to about 480 km above the surface but the density of the air decreases with height. Most of the mass of the atmosphere is within 16 km of the Earth's surface.

FIG 9.1.3 The Earth's atmosphere

Although this sounds quite thick, it is only a thin layer of gases compared to the size of the Earth. Nevertheless it contains two gases, carbon dioxide and oxygen, which complement water in also being essential for life on the planet.

Once again, supplies of these gases are maintained by natural processes. The carbon cycle provides a means of converting carbon dioxide from the air into chemicals which are essential to plants and animals. These are subsequently replaced in nature by the decay of waste and dead organic materials.

Fun fact

Before plants able to carry out photosynthesis evolved, there was no free oxygen in the atmosphere. This means that animals could not have evolved without the help of plants.

Sources and uses of water

We are learning how to:

- describe where water comes from
- identify safe sources of water
- describe how people use water.

Sources of water >>>

Some people who live in towns and cities are lucky. They have water supplied to their homes through pipes.

In rural communities people are more likely to obtain their water from communal standpipes or **wells**.

A well is a hole bored into the ground to reach water that collects below the surface. The water is pumped to the surface using an electric pump. This can be driven by **solar cells** which convert sunlight into electricity.

Where there is no supply of fresh drinking water people have no choice but to obtain the water they need from rivers. This is not a satisfactory practice.

Even when water looks clear it might be contaminated by products of human or animal waste. It may also contain harmful chemicals from fields or organisms that cause disease like typhoid and cholera.

FIG 9.2.1 A water supply to someone's home

Uses of water

The amount of water used each day by people in Jamaica varies according to a number of factors such as whether they live in towns or in the country, and whether they use water in their work, e.g. to make concrete or to irrigate fields.

Apart from drinking people typically use water in other ways including:

- Personal hygiene
- Washing clothes
- Cleaning homes
- Preparing and cooking food
- Providing drinking water for livestock and pets
- Watering the plants in pots and gardens.

The ways in which people use water can be placed into two categories:

- Essential uses of water – these are necessary for keeping healthy and maintaining a clean safe environment.
- Non-essential uses of water – these may give us pleasure but they are not necessary to our health and well-being.

An adult needs to consume between 2 and 3 litres of water each day to remain healthy and a child proportionally less. Some of this is provided by foods like fruits which contain lots of water.

It's nice to have a clean car to drive about in but this is not essential for a person's health or well-being.

Activity 9.2.1

A survey of how my family uses water in our home

You don't need any equipment for this activity.

Here is what you should do:

1. Make a list of ten ways you use water at home. Put them in a table like Table 9.2.1.

2. Alongside each, say whether you think it is essential or non-essential.

3. Alongside each, estimate the total amount of water for that use each day in your home. The following information will help you:
   - Taking a bath uses 80 l
   - Taking a shower uses 40 l
   - Flushing a toilet uses 8 l
   - Running a washing machine uses 65 l per load
   - A bucket contains 8 l

Use of water	Essential or non-essential?	Total amount of water used

TABLE 9.2.1

4. Make a list of the ways in which your family could reduce its water usage.

Check your understanding

1. You are out on a walk in the country on a warm day and see what looks to be clean cool water in a pond. Is this water safe to drink? Explain your answer.

2. Decide whether each of the following is an essential or a non-essential use of water in a period of drought.

   a) Washing the family car to remove the dust from it.

   b) Cleaning your teeth and rinsing out your mouth.

   c) Drinking when you become thirsty.

   d) Watering the flower garden with a hosepipe.

   e) Hosing yourself to keep cool on a hot day.

Fun fact

Before electric pumps, people used to raise water from a well in a bucket using a simple machine called a windlass.

FIG 9.2.2 A windlass

Key terms

well hole bored or dug from the surface to water collected beneath

solar cell device that transforms energy from sunlight into electricity

Properties of water

We are learning how to:

- describe the properties of water
- test for the presence of water.

States

Water is a colourless liquid at room temperature but it can be converted to a solid or to a gas by changing its temperature.

The freezing point of water is 0°C under normal atmospheric pressure. At this temperature liquid water turns to solid water, which we call ice. At the same temperature ice turns to water.

The boiling point of water is 100°C at normal atmospheric pressure. At this temperature liquid water turns to gaseous water, which we call steam. At the same temperature steam turns to water.

Notice that steam is actually colourless. The white 'smoke' we see is composed of tiny droplets of water which have condensed in the cooler air.

Testing for water

A simple test for water uses blue cobalt chloride paper.

When dipped into water blue cobalt chloride paper turns pink. This is a test for the presence of water but not its purity. Any aqueous solution will give a positive result.

Copper(II) sulfate is a blue solid but when heated, water is driven off and white anhydrous copper(II) sulfate is formed. This is a reversible reaction so anhydrous copper(II) sulfate can also be used to test for water:

$$- \text{water}$$
$$\text{blue copper(II) sulfate} \rightleftharpoons \text{white anhydrous copper(II) sulfate}$$
$$+ \text{water}$$

Once again, this is a test for the presence of water but not its purity.

FIG 9.3.1 Water freezes at 0°C

FIG 9.3.2 Water boils at 100°C

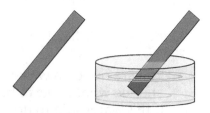

FIG 9.3.3 Water turns blue cobalt chloride paper pink

FIG 9.3.4 Water turns white anhydrous copper(II) sulfate blue

Activity 9.3.1

Investigating other properties of water

Here is what you need:

- Microscope slides × 2
- Beaker
- Capillary tubes of different inner diameters
- Metal pin
- Tweezers
- Soap solution
- Water
- Coloured ink
- Dropper pipette.

Here is what you should do:

1. Place a microscope slide on the bench and put a couple of drops of water onto it. Carefully place a second microscope slide on top of the first.

2. Try and pull thee slides apart without sliding them over each other. Comment on your observations.

3. Half-fill a beaker with water and add a few drops of ink to give a coloured solution.

4. Stand capillary tubes in the beaker and leave them for a few minutes.

5. Comment on the height of water in each of the capillary tubes.

6. Half-fill a beaker with water.

7. Using tweezers carefully lower a metal pin so that it sits on the surface of the water.

8. Add a few drops of soap solution to the water, being careful not to disturb the pin, and observe what happens.

Key terms

cohesion force of attraction between water molecules

surface tension ability to support a small object on the surface of water as a result of cohesive forces between molecules of water

adhesion force of attraction between water molecules and other materials

capillarity movement of water up the inside of a capillary tube

Water molecules are attracted to each other and also the walls of their container. This results in some interesting properties.

Cohesion occurs as a result of attraction between molecules of water. This creates **surface tension** at the surface of water in a container. The attractive forces are sufficiently strong to support a small object like a pin. When soap is added the surface tension is reduced.

Adhesion occurs as a result of attraction between molecules of water and other materials. When water is placed between two microscope slides the adhesion forces must be overcome in order to separate them.

When a capillary tube is placed in a beaker of water, adhesion draws water up the inside of the tube so that the level is higher than the surface of the water. This behaviour is called **capillarity**.

Check your understanding

1. Under normal atmospheric pressure at what temperature does:

   a) ice turn to water?

   b) water turn to steam?

2. a) A colourless liquid turns blue cobalt chloride paper red. Does this mean that the liquid is pure water?

   b) Give details of a second test you could carry out to show the presence of water.

Fun fact

The boiling point of water decreases with height above the ground. On top of Mount Everest water boils at 71 °C.

Water purification and conservation

We are learning how to:

- describe the different stages in water purification
- reduce water consumption.

Purifying drinking water ≫

The water for drinking may be taken directly from rivers, or stored in **reservoirs** to ensure continuity of supply during periods with no rain. The Mona Reservoir provides water for Kingston.

In order to make sure that water is safe to drink it must pass through a water treatment plant.

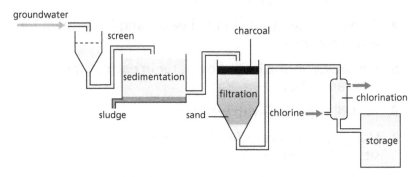

FIG 9.4.1 Purifying drinking water

There are several stages to water treatment:

- Water first passes through a screen which removes large debris, such as pieces of tree.

- From there it goes into a **settling tank** where large particles are allowed to settle out, forming sludge at the bottom of the tank.

- Finer particles are removed by **filtration** through a bed of sand.

- In order to ensure no harmful organisms remain in the water, it is **chlorinated** by forcing chlorine gas through it.

- The treated water is held in storage tanks until it is used.

Fun fact

A single leaking tap that drips once every 10 seconds loses around 1 dm³ of fresh water each day. There are around one million households in Jamaica. If each household had one dripping tap that is a lot of fresh water wasted.

Activity 9.4.1

Making a filter bed

Here is what you need:

- Muddy water in a beaker
- Empty 330 ml plastic bottle
- Stand and clamp
- Fine gravel
- Sand
- Fine plastic net or similar material.

1. Remove the screw cap from the bottle and cut the base off the bottle.

2. Support the bottle upside down using a stand and clamp.

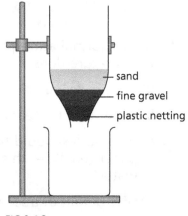

FIG 9.4.2

3. Place some plastic net in the neck of the bottle.

4. Add gravel to a depth of about 5 cm and then add a similar depth of sand (Fig 9.4.2). You now have a filter bed.

5. Place a clean beaker beneath the filter bed.

6. Pour the muddy water into the filter bed and leave it to pass through the filter (Fig 9.4.3).

7. How does the water in the beaker look compared to the muddy water? Do you think this water is safe to drink?

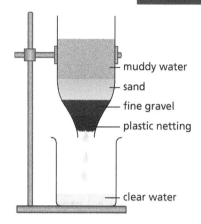

FIG 9.4.3

Conserving water

Recent studies show that, as with some other Caribbean countries, Jamaica will not be able to keep up with the rising demand for fresh water in the future. A simple way of conserving water is to find more efficient ways of using the water we have.

Each household could reduce the amount of fresh water they use without compromising on health and hygiene simply by modifying their lifestyle. Here are some examples you might recognise:

- Turn off the tap while cleaning your teeth and turn it on again to wash your toothbrush. This uses less water than leaving the tap running.

- Pour the water used to clean fruits and vegetables onto plants in the garden. You save water because you don't need to water the plants so often with a hosepipe.

- Repair leaking taps. Every drip is an unnecessary loss of fresh water.

- Install water butts to collect rainwater and use this to water plants. This reduces the amount of fresh water you need to use on your pots and garden.

Check your understanding

1. **a)** In which order do the following take place during water purification?

    chlorination screening filtration sedimentation

   b) Which of these processes ensures that the water contains no harmful microorganisms?

2. Suggest three ways of conserving water which are not mentioned in the text. These should not compromise health or hygiene.

Key terms

reservoir artificial lake used to store water

settling tank where particles are allowed to settle out from water during purification

filtration method of separating solid and liquid

chlorination killing microorganisms in water by bubbling chlorine gas through it

The water cycle

We are learning how to:

- understand why the world doesn't run out of fresh water
- describe the different processes in the water cycle.

The water cycle »»

Water is an essential requirement in our lives. We use many millions of cubic decimetres every day yet fresh water doesn't run out. Supplies of fresh water are replenished by a series of natural processes we call the water cycle.

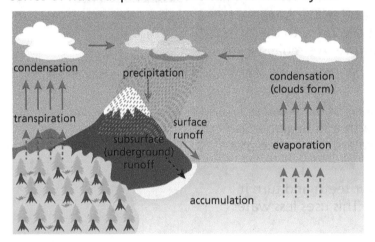

FIG 9.5.1 The water cycle

The energy needed to drive this cycle is provided by the Sun. In the warm regions of the world, **heat radiation** from the Sun causes massive amounts of water to **evaporate** into the atmosphere which condenses to form clouds of tiny water droplets. This can be described as an **endothermic** process because energy is taken in.

Water is also released into the atmosphere through **transpiration**. This is the process by which plants lose water, mostly through their leaves.

The clouds are carried around the Earth by the prevailing winds. As they pass into cooler regions, the air cannot carry so much water vapour. Some water vapour is therefore lost as rain, or in very cold regions, as snow. This is called **precipitation**.

Rain falling on the land flows across soil and rocks, and eventually accumulates in lakes and rivers. Sometimes artificial lakes called reservoirs are built to store water because it may not rain throughout the year. Water stored in the rainy season can be used in the dry season.

Fun fact

In some countries the water cycle does not produce sufficient rain to satisfy the demand for fresh water. Additional fresh water is made from sea water. This process is called desalination.

FIG 9.5.2 Pure water falls to the ground as rain or snow

Sometimes soluble minerals dissolve in the water as it flows over rocks above and below the ground. The mineral water may accumulate in **natural springs**.

People often buy spring water that comes from places like Portland, or from the Blue Hole Mineral Spring in Negril.

FIG 9.5.3 Spring water

Activity 9.5.1

Building a small-scale water cycle

Here is what you need:

- Beaker × 2 – one large and one small
- Elastic band – large
- Small stone or glass marble
- Polythene sheet
- Salt water.

Here is what you should do:

1. Pour salt water into a large beaker to a depth of about 2 cm.

2. Thoroughly wash and dry a small beaker. You are going to drink from this beaker later in the activity so it is essential that it is clean.

3. Place the small beaker inside at the centre of the large beaker.

4. Place the polythene sheet over the top of the large beaker and put the small stone in the centre (Fig 9.5.4). Allow the sheet to sag at the centre so the small stone is directly above the small beaker.

5. Hold the polythene sheet in place with an elastic band.

6. Place the apparatus on a sunny windowsill for a day.

7. After a day, remove the small beaker, wipe any salt water from the outside and taste the contents.

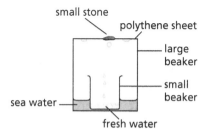

FIG 9.5.4

Key terms

heat radiation heat transferred as rays

evaporate change from liquid to vapour below the boiling point of the liquid

endothermic taking in heat

transpiration process by which plants lose water, mainly through their leaves

precipitation water falling to the ground as rain or snow

natural springs places where rain water flows out of rocks

Check your understanding

1. a) What provides the energy that drives the water cycle?

   b) What happens when water evaporates?

   c) Rain often forms from sea water but it doesn't taste salty. Explain why.

   d) What form does precipitation take in Jamaica?

   e) How is mineral water formed from rain?

Composition of the atmosphere

We are learning how to:

- name the gases in the atmosphere
- describe how to estimate the percentage of oxygen in air.

Composition of air >>>

Air is a mixture of gases that surrounds the Earth. Air is a mixture and not a compound because it does not have a fixed composition and the gases are not chemically combined together.

Typical percentages of the gases found in dry air are given in Table 9.6.1.

Gas	Percentage of dry air
Nitrogen	78
Oxygen	21
Argon	1
Carbon dioxide	0.04

TABLE 9.6.1

Activity 9.6.1

Estimating the percentage of oxygen in air

When iron is exposed to damp air it reacts with oxygen to form **rust**. You are going to use this reaction to estimate the percentage of oxygen in air.

Here is what you need:

- Beaker
- Test tube
- Stand and clamp
- Iron wool
- Stirring rod
- Ruler.

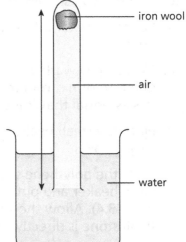

FIG 9.6.1

Here is what you should do:

1. Use a stirring rod to push a small ball of iron wool into the bottom of a test tube. It must be tight enough that it doesn't fall out when the test tube is turned upside down.
2. Stand the inverted test tube in a beaker of water so that the open end is at least 2 cm below the surface (Fig 9.6.1).
3. Clamp the test tube in place.
4. Measure the height of the column of air in the inverted test tube.
5. Leave the apparatus for one week but top up the beaker with water regularly so that the surface of the water is well above the open end of the test tube.
6. After one week:
   a) observe the appearance of the iron wool
   b) observe the level of water in the test tube
   c) measure the height of the column of air in the test tube.
7. Explain your observations.
8. Use your observations to estimate the percentage of oxygen in air.

Maintaining the composition of the air

Oxygen is very reactive and takes part in a number of everyday processes.

The following processes remove oxygen from the air on a large scale.

- Oxygen is needed for the **combustion** of fuels like natural gas and petrol:

    fuel + oxygen = carbon dioxide + water + energy

- Oxygen is needed by all living things, plants and animals. It is used in the process of **respiration**:

    glucose + oxygen = carbon dioxide + water + energy

- Oxygen is also involved in the corrosion of some metals such as iron.

In the presence of air and moisture, iron is oxidised to iron oxide which is better known as rust:

    iron + oxygen = iron oxide

Because all of these processes take place continually, how is it that the supply of oxygen from the air did not run out millions of years ago? The answer is that there is another process which replenishes the Earth's supplies of oxygen. This process is called **photosynthesis**:

$$\text{carbon dioxide + water} \xrightarrow[\text{chlorophyll}]{\text{sunlight}} \text{glucose + oxygen}$$

In addition to maintaining the level of oxygen in the atmosphere, photosynthesis also removes the carbon dioxide produced by combustion and respiration.

FIG 9.6.2 Green plants carry out photosynthesis during the day

Key terms

air a mixture of gases that forms the atmosphere

rust a type of iron oxide formed when iron reacts with oxygen in the air

combustion another term for burning

respiration process that uses oxygen to release energy in the cells of all living organisms

photosynthesis process in green plants that makes food and oxygen

Check your understanding

1. Fig 9.6.3 shows apparatus used in an activity to investigate air. Hot copper reacts with oxygen to form black copper(II) oxide.

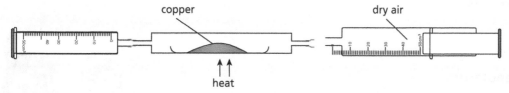

FIG 9.6.3

The copper is heated and then air is passed from one syringe to the other until a constant volume of air is obtained. The apparatus is then left to cool.

a) What is the aim of this activity?

b) Which gas in air reacts with hot copper?

c) What evidence will there be at the end of the activity that a chemical reaction has taken place?

d) Predict the total volume of gas in the syringes at the end of the activity.

How the gases in air are used

We are learning how to:

- describe how the gases in air are used
- carry out a test for oxygen and carbon dioxide.

Gases in air ▶▶▶

The four main gases in air in order of composition are nitrogen, oxygen, argon and carbon dioxide.

Nitrogen

Nitrogen is a colourless and odourless gas, which does not readily take part in chemical reactions. This **inert** nature gives nitrogen a number of important uses.

Nitrogen is often used in order to exclude oxygen, which is a more reactive gas.

- Some foods are packaged in an atmosphere of nitrogen in order to extend their **shelf life**. The bacteria that cause the food to decay would be far more active in the presence of oxygen.

- **Welding** metal is sometimes carried out in an atmosphere of nitrogen. At the very high temperatures involved, metals would react with atmospheric oxygen and weaken the weld.

- Nitrogen is sometimes used in cheap filament lamps to prevent the reaction of the filament with oxygen.

Nitrogen has one very important reaction in the manufacture of a chemical compound called ammonia. Ammonia is then used to make nitrogenous **fertilisers** like ammonium nitrate.

FIG 9.7.1 Ammonium nitrate is an important chemical fertiliser

Oxygen

The main industrial uses of oxygen are:

- steel manufacture
- burning with fuel gases to produce flames hot enough to melt and cut through steel.

Oxygen is also a colourless and odourless gas.

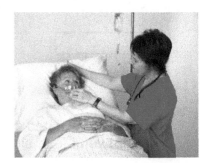

FIG 9.7.2 Oxygen helps people with breathing difficulties

Activity 9.7.1

Preparing and testing for oxygen

Hydrogen peroxide decomposes in the presence of manganese(IV) oxide to release oxygen. You are going to use this reaction to make the gas.

Here is what you need:

- Test tube
- Manganese(IV) dioxide
- Hydrogen peroxide 20 volume
- Wooden splint.

Here is what you should do:

1. Add hydrogen peroxide to a test tube to a depth of about 2 cm.

2. Place a small amount of manganese(IV) dioxide into the test tube.

3. Light a wooden splint and then blow it out so that it is still smouldering and glowing red (Fig 9.7.3).

4. Place this in the top of the test tube.

5. Observe what happens.

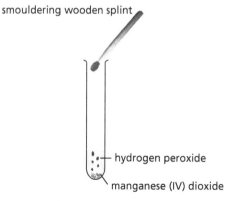

smouldering wooden splint

hydrogen peroxide

manganese (IV) dioxide

FIG 9.7.3

Argon

Argon is used in a similar way to nitrogen where oxygen is to be excluded. In a filament lamp the tungsten filament glows white hot at a temperature of between 2 000 °C and 3 000 °C. At such a high temperature tungsten would react with any oxygen present and the filament would soon be destroyed. To prevent this, the bulb is filled with argon.

Carbon dioxide

Carbon dioxide gas is a colourless and odourless gas that is denser than air. It is very soluble in water and dissolves to form a weak acid called carbonic acid.

Carbon dioxide is essential for the process of photosynthesis but also has other important uses.

- Carbon dioxide gas does not support combustion and is denser than air so it can be used to extinguish fires.

- Carbon dioxide is not toxic and dissolves readily in water. It is used to make drinks fizzy.

FIG 9.7.4 Argon is an inert gas used in light bulbs

Check your understanding

1. State one use for each of the four main gases in air.

2. Briefly describe the test for the presence of:

   a) oxygen gas
   b) carbon dioxide gas.

Key terms

inert having few chemical reactions

shelf life amount of time food can remain on display in a shop and still be fit to eat

welding joining pieces of metal together using heat

fertiliser substance that assists plant growth

The carbon cycle

We are learning how to:

- describe how carbon exists as carbon dioxide in the atmosphere, as carbonates in rocks and as organic chemicals in living things
- understand how carbon is continually recycled by natural processes.

Carbon cycle »»

Modern society is very dependent on carbon-based **fuels** like coal, natural gas and fuels derived from crude oil to provide energy. Energy cannot, however, be created nor destroyed, so where does it come from?

With the exception of nuclear fuels, like uranium and plutonium, the energy in fuels comes from the Sun through the process of photosynthesis. The glucose formed during photosynthesis might be stored in the plant as starch or it might be converted by the plant into other essential chemicals.

Glucose is the energy source for plants and for animals that eat the plants. During the process of respiration energy is released in cells.

In some ways photosynthesis and respiration can be considered as chemical opposites. These two processes together provide a means of transferring energy from the Sun into the cells of living things.

There are also other processes like combustion and decay that release carbon, in the form of carbon dioxide, back into the environment. The relationship between all of these processes can be shown as a carbon cycle.

The carbon contained in complex chemicals which make up the structure of a plant or an animal will eventually end up back in the atmosphere. The bacteria which bring about the decomposition of organic waste, or of the dead plant or animal, are an essential part of the recycling process.

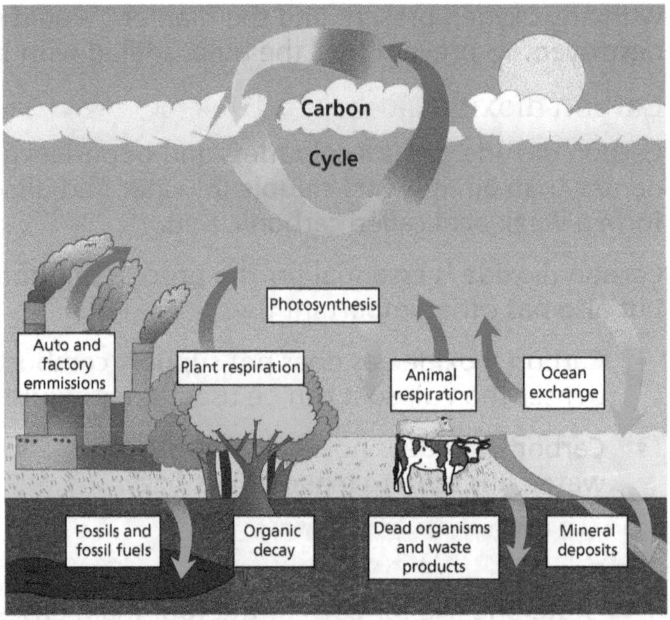

The processes that remove oxygen from the air and release carbon dioxide, like respiration, combustion and decomposition, are exactly balanced by the process of photosynthesis which removes carbon dioxide and releases oxygen. The composition of the atmosphere therefore stays more or less constant.

The sea also plays an important part in keeping the composition of the atmosphere constant by absorbing carbon dioxide from the air to form a very weak acid called **carbonic acid**.

FIG 9.8.1 The carbon cycle

In the sea, carbonic acid slowly reacts with other substances to form an insoluble chemical called calcium **carbonate**. This is the chemical from which marine animals make their shells, and from which huge coral reefs are formed.

Over millions of years, some of the calcium carbonate may form rocks like limestone. There are large deposits of **limestone** in Jamaica and it is widely used for building and in the manufacture of cement.

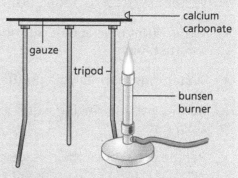

Activity 9.8.1

Thermal decomposition of calcium carbonate

FIG 9.8.2 Sea shells are composed of calcium carbonate

Here is what you need:

- Tripod
- Gauze
- Bunsen burner
- Tongs
- Boiling tube × 2
- Drinking straw
- Filter funnel
- Filter paper
- Limestone.

FIG 9.8.3

Here is what you should do:

1. Place a small piece of limestone on the edge of a gauze that is supported on a tripod (Fig 9.8.3).

2. Heat the limestone strongly for 10 minutes.

3. Allow the limestone to cool.

4. Using tongs, place the heated limestone in a boiling tube.

5. Add a few drops of water to the boiling tube. Hold the bottom of the tube and feel if the contents get warmer or cooler.

6. Half-fill the boiling tube containing the heated limestone with water.

7. Mix the contents of the boiling tube and then pour them into a filter paper and funnel. Allow the filtrate to collect in the second boiling tube.

8. Place a drinking straw into the filtrate and gently blow until you notice a change taking place.

9. What gas in your exhaled breath causes this change?

Key terms

fuel substance that releases heat when burned

carbonic acid formed when carbon dioxide dissolves in water

carbonate chemical containing carbon

limestone type of rock containing mainly calcium carbonate

Check your understanding

1. Describe the processes by which a molecule of carbon dioxide in the air might become:

   a) part of a plant

   b) part of an animal

   c) part of a rock.

Review of Water and the Earth's atmosphere

- Around 70% of the Earth's surface is covered in water.

- Over 99% of the water on Earth is contained in the seas.

- Sources of fresh water include rivers, lakes, wells and springs.

- The uses of fresh water can be divided into those which are essential, like drinking and washing, and those which are non-essential, like washing a car.

- Water may exist as a solid (ice), a liquid (water) and a gas (steam).

- Water turns blue cobalt chloride paper pink and turns white anhydrous copper(II) sulfate blue.

- Water must normally be purified before it is fit to drink.

- Purification of drinking water includes screening, sedimentation, filtration and chlorination.

- People can reduce their usage of fresh water by simple actions like turning a tap off when it is not being used and repairing dripping taps.

- Supplies of fresh water on the Earth are continually replenished by the water cycle.

- The water cycle involves a sequence of natural processes.

- The Sun provides the energy needed for the water cycle.

- The atmosphere is a layer of gases surrounding the Earth.

- The atmosphere is a mixture of gases called air.

- Dry air consists of about 78% nitrogen, 21% oxygen, 1% argon and 0.04% carbon dioxide.

- Oxygen is used up during respiration, combustion and rusting.

- Oxygen is produced during photosynthesis.

- Nitrogen is relatively inert and is used to provide an inert atmosphere in applications like extending the shelf life of food and in welding. It is also used in the manufacture of nitrogenous fertilisers like ammonium nitrate.

- Oxygen is used in hospitals where people have trouble breathing only air, and to produce hot flames for cutting steel.

- Oxygen rekindles a glowing splint of wood.

- Argon provides an inert atmosphere in filament lamps.

- Carbon dioxide is needed for photosynthesis.

- Carbon dioxide turns limewater milky.

- The carbon cycle is a sequence of natural processes in which carbon is taken from the air in the form of carbon dioxide during photosynthesis to form food and other chemicals in plants. The chemicals may be eaten by animals. Eventually, as waste or dead organic material, the chemicals containing carbon decompose, releasing the carbon dioxide back into the air.

- Carbon dioxide is absorbed into the seas and eventually forms calcium carbonate, which is used to form the shells of marine organisms.

Review questions on Water and the Earth's atmosphere

1. **a)** Name three sources of fresh water.

   b) Is the water from each of these sources fit to drink? Explain your answers.

2. **a)** Give three examples each of:

   i) essential uses of water   **ii)** non-essential uses of water.

   b) Give an example of how a person could conserve water by making a small change to the way they do things.

3. **a)** In what state is water at a temperature of:

   i) 110°C?   **ii)** −10°C?   **iii)** 25°C?

   b) State two tests you could use to demonstrate that a liquid contains water.

   c) Do these tests indicate that the water is pure? Explain your answer.

4. Fig 9.RQ.1 shows what happened when some tubes of different internal diameters were stood in a beaker of water.

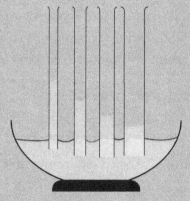

FIG 9.RQ.1

   a) What pattern is there between the internal diameter of a tube and the height of water in it?

   b) Explain why water behaves in this way.

5. Briefly explain the purpose of each of the following steps in water purification.

   a) Screening   **b)** Sedimentation

   c) Filtration   **d)** Chlorination

6. Draw a labelled diagram of the water cycle.

7. The following diagrams show what happened when an excess of the element phosphorus was burned under a bell jar of air.

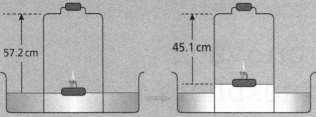

57.2 cm   45.1 cm

FIG 9.RQ.2

   a) Which gas does phosphorus react with when it burns?

   b) Use the data in the diagram to calculate the percentage by volume of this gas in air.

8. a) i) Which gas from air is needed by plants to make food?

   ii) Describe a test for this gas.

   b) i) Which gas from air is used in filament lamps?

   ii) What property of this gas makes it suitable for this purpose?

   c) i) Which gas from air is sometimes given to patients in hospital when they are finding it difficult to breathe?

   ii) Describe a test for this gas.

9. a) In what form does carbon exist in air?

   b) In what form does carbon exist in sea shells?

   c) In what form does carbon exist in food made by green plants?

   d) In what form may carbon be stored for millions of years in the ground?

   e) What happens to the carbon contained in organic material when it rots?

10. a) What is the difference between cohesion and adhesion?

   b) Fig 9.RQ.3 shows an insect called a pond skater.

   Explain how a pond skater is able to walk on water.

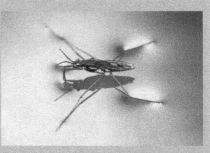

FIG 9.RQ.3

11. In 1894 the Scottish chemist Sir William Ramsay noticed that dry nitrogen made by a chemical reaction has a density of 1.198 kg m^{-3} while nitrogen made by eliminating carbon dioxide and oxygen from air has a density of 1.204 kg m^{-3}.

    a) Why was it important that the samples of nitrogen were dry?

    b) Explain why there was a difference in the density of nitrogen made by these two different methods.

    c) Which element was discovered as a result of Ramsay's careful observation?

12. Figure 9.RQ.4 shows the apparatus used by Lavoisier to investigate air. He placed some mercury in a long-necked retort flask on a charcoal furnace. The neck of the flask led into a bell jar sat in a trough of mercury.

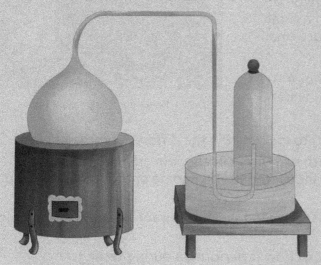

FIG 9.RQ.4

After heating the mercury in the flask for some days he noticed that red particles appeared on its surface but after a while these did not seem to increase in quantity. He then put the fire out and, by measurement, found that about one-sixth of the air had disappeared. He called the gas remaining in the bell jar 'azote' or nitrogen.

    a) Suggest the chemical name of the red particles Lavoisier noticed on the surface of the mercury.

    b) How was Lavoisier sure that all of the oxygen in the air had reacted?

    c) Give the percentage by volume of oxygen and nitrogen in air according to Lavoisier's measurements.

Water and the Earth's atmosphere

· ·

Often fresh water may appear to be clean but it is not considered safe to drink until it has been purified.

There are several stages in water purification. Your task is to make a working model of a water purification plant which has these four stages of purification. Your model will be the prototype of a water purification system that people in a rural community would be able to use to produce small volumes of drinking water.

FIG 9.SIP.1 Water purification plant

1. You are going to work in groups of 3 or 4 to make models of the four parts of a water purification plant. The tasks are:

   • To revise how water is purified and to look at how the different processes work.

   • To devise and build apparatus which will carry out the four stages of purification using readily available containers and materials. Take some photographs as you do this to illustrate your report.

   • To test how well your model works. You can take photographs or make a short video of this.

   • To modify your model on the basis of how it performed during testing.

   • To compile a report on the construction of your water purification plant including a demonstration of the different stages. At the end of your report you should be prepared to discuss how your design could be scaled up to make larger volumes of water.

   a) Look back through the first part of the unit and particularly lesson 9.4, which is about the purification of water. Look carefully at the different stages which include:

   • Screening to remove large debris.

   • Sedimentation to allow large particles to settle.

   • Filtration to remove smaller particles.

   • Chlorination to kill harmful microorganisms.

   b) At this stage focus on each of the different processes rather than worrying how they are to be connected together. You need to think in detail about each one and design a suitable apparatus. The following might give you some ideas.

   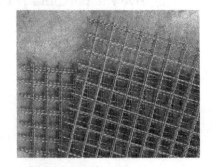

   FIG 9.SIP.2 Metal gauze

   Metal or plastic gauze could be used to screen large objects out of water. How big do the holes in the gauze need to be? If you use the gauze horizontally it will soon be covered in debris and stop working. Is it possible to mount it at an angle so the water flows through while the debris falls to the side?

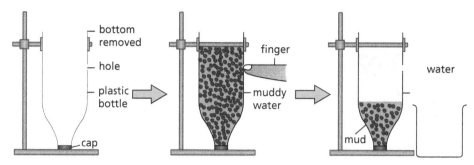

FIG 9.SIP.3 Sedimentation

A simple sedimentation tank can be made from an empty water bottle. Can you start with this design and improve on it?

You made a filter bed in Activity 9.4.1. You can include this in your design in your water purification apparatus or you can modify it if you think it necessary.

Chlorine is a poisonous gas so it needs to be handled carefully. If you decide to try and chlorinate the water using this gas directly your teacher will help you to decide on a suitable chemical reaction for producing small amounts.

FIG 9.SIP.4 Purification tablets

Instead you might prefer to pass your clear water down a column containing water purification tablets. How long does the water need to be in contact with the tablets to be purified?

c) Test each part of your apparatus separately to ensure everything works as it should. When you are satisfied, it is time to test the whole purification process.

Start with a sample of pond water or something similar, and pass it through each stage.

To test whether the water is safe to drink you are going to have to check if there are any microorganisms in it. You can do this using agar gel in a Petri dish.

FIG 9.SIP.5 Testing water for purity

Smear some of your purified water onto freshly made agar gel in a Petri dish. Place the top on the Petri dish and tape around the sides to stop any organisms entering from the air. Place the dish somewhere warm for a few days to see if any colonies of microorganisms grow on it. Use your results to determine how effective your water purification has been. Take a photograph of the agar gel to support your report.

d) If your apparatus has not produced purified water, or as a result of using it, you can see a better way to do things, modify your design and retest it.

If you have sufficient time you might experiment to see if you can connect your apparatus together in some way to make a continuous water purification process.

e) Compile a report on what you have done. Your report should contain details of how you constructed the different parts of your apparatus and how easy they were to use. Illustrate your account with some photographs.

Discuss how well your apparatus performed during the trial and show a picture of the agar gel plate used to test how pure the water was.

Complete your report by discussing how the different stages in your process could be modified to make a continuous process. Also discuss how the apparatus might be scaled up to produce larger volumes of fresh water.

Index

Note: Page numbers followed by f or t represent figures or tables respectively.

A

absorption, 65
acceleration, 144, 145
acids, 94, 95
    amino, 66
activity-related energy, 70, 71
adhesion, 209
air, 101, 156
    composition of, 214–215, 214f
    exhalation, 157
    gases in, 216–217
air resistance, 128, 129, 132, 133
alimentary canal, models of, 86–87, 86f, 87f
alkali metals, 46, 47
alloy, 102, 102f, 103
aluminium, 98f, 99
alveoli, 154, 155
    diffusion of gases in, 156, 156f
amino acids, 66
ammonia, 180, 181, 216
    diffusion of, 53, 53f
ammonium nitrate, 216, 216f
Angostura Bitters distillery, 113
animals, 26–27, 26f, 38
anomalous values, 11, 11f
Apollo 11 mission, 185, 185f
aqueous solution, 104, 105
Archimedes Principle, 134, 135
argon, 217, 217f
assimilation, 65
asteroid belt, 174f, 175, 176, 176t
asteroids, 194, 195
atmosphere, 178, 179
atomic number, 48, 49, 50, 51
    and Periodic Table, 50, 50f
atomic symbols, 42, 42t, 43
atoms, 40–41, 40f, 42, 43
    of carbon, 50, 50f
    nucleus, 48
    overview, 48
    sub-atomic particles, 48–49, 48f, 49t, 50–51
average speed, 138, 139
axis of rotation, 180, 181

B

balanced diet, 68, 69, 71
balanced forces, 123, 123f, 126, 127
bar charts, 8, 8f, 9
basal metabolic rate (BMR), 70, 71
base, 94, 95
Benedict's reagent, 78, 79
Big Bang Theory, 198, 199
biological magnification, 32, 33
Biuret reagent, 80, 81
blood, 100
Blue Planet, 204, 204f
BMR. see basal metabolic rate (BMR)
boiling points, 113
    of water, 208, 208f, 209
Bolt, Usain, 139
breathing, 152
    exhaling, 158, 158f, 159
    inhaling, 158, 158f, 159
breathing rate, 153, 162–163, 162f, 163f
    exercise and, 164–165
bronchioles, 154, 155
bronchus, 154, 154f, 155
bronze, 103, 103f
buoyancy force, 134, 134f, 135
burning, 97

C

calcium carbonate, 219, 219f
calorie, 69
calorimeter, 71
carbohydrates, 20, 21, 66, 67
carbon, 97
    atoms of, 50, 50f
    isotopes of, 51, 51f
carbonate, 219
carbon cycle, 218–219, 218f

carbon dioxide, 156, 157, 160, 178, 217
carbonic acid, 217, 218–219
carnivores, 19, 27, 27f
cells
    heart muscle, 161, 161f
    respiration in, 24, 25, 152, 160–161, 160f
    skeletal muscle, 161, 161f
Ceres, 175
cheetah, 137
chemical changes, 76, 77, 90, 120–121
    burning, 97
    colour change, 95
    effervescence, 94
    physical changes vs., 96–99
chemical compounds, 104
chemical reaction, 99
chlorination, 210, 211
chlorine, 225
chlorophyll, 18, 19, 22, 23, 23f
chromatography, 114–115, 115f
clusters, 197, 197f
cohesion, 209
combustion, 215
comets, 194–195, 194f
compounds, 104, 104t, 106, 107
concentration, 52, 53
concentration, of gas, 156, 157
concentration gradient, 52–53
conductors, 44, 45
constants, 12
consumers, 19, 26, 26f, 30, 31
    primary, 26, 27
    secondary, 26, 27
control variables, 12, 13
copper(II) sulfate, 208
crude oil, 102, 102f, 103

D
Dalton, John, 48, 48f
data, 6, 7
    anomalous value, 11, 11f
    patterns, 10, 10f, 11
    presentation of, 8–9
DDT (dichlorodiphenyltrichloroethane), 32, 33
deceleration, 144
decomposers, 19, 27

density, 134–135, 134f, 177
    relative, 150
dependent variables, 12, 13
diabetes, testing to diagnose, 79, 79f
diaphragm, 158, 159
diet
    balanced, 68, 69, 71
    defined, 68, 69
dietary requirements, 88–89, 88f
differentially permeable membrane, 54, 54f, 55
diffusion, 41, 52–53
    of gases in alveoli, 156, 156f
    rate of, 53
digestion, 65
    process of, 76–77
digestive system, 65, 72–75, 72f
displacement, 124
    and velocity, 142–143
distance
    relationship between speed and time,
        138–139
    in space, 198–199
distance-time graphs, 140–141, 140f, 141f
distillate, 111
distillation, 111
    of liquid mixtures, 112–113
distilled water, 91, 91f, 100, 101
duralumin, 102, 102f, 103

E
Earth, 143, 143f, 177, 178, 178f, 182, 182f
    atmosphere, water and, 204–205, 204f,
        205f, 224–225
    gravity, 182–183, 182f
eclipse
    lunar, 192–193, 193f
    solar, 190–191, 190f
ecosystem, 22, 23
effervescence, 94, 95
egestion, 65
electricity, conductor of, 44, 45
electrons, 48, 49
elements, 40, 40f
    Group 17, 47, 47f
    non-metallic, 44, 44f
    periodic table of, 42–43, 43f
    transition, 47, 47f

elliptical galaxy, 196, 196f, 197
endothermic, 212, 213
energy
    activity-related, 70, 71
    from food, 153, 166–167, 166f
    and friction, 131
    kinetic, 41, 41f, 52
    needs, 70–71
    and photosynthesis, 38–39
epiglottis, 75
equal, in magnitude, 125, 125f
equilibrium, 54, 55
escape velocity, 143
ethanol, 80, 81, 113
evaporate/evaporation, 110, 212, 213
excretion, 65
exercise, and breathing rate, 164–165
exhaling, 158, 158f, 159
exothermic reactions, 106, 107
extended light source, 188, 188f, 189

F
fats, 66, 67
    testing for, 80, 80f
fertilisers, 216, 217
fibre, 64, 67
filtrate, 109
filtration, 109, 109f, 210, 211
float, 134–135, 134f
fluorine, 47
food, 64, 64f
    breaking down, 76–77
    chains (see food chains)
    energy content of, 69
    energy from, 153, 166–167, 166f
    groups (see food groups)
    tests, 65, 78–81
food calorimeter, 166, 166f
food chains, 19f, 24, 25, 28–29, 32, 33, 38
    impact of human activity on, 32–33
food groups
    'glow' foods, 67, 71
    'go' foods, 66, 70
    'grow' foods, 66, 66f, 70
food webs, 30–31, 30f, 38

force diagrams, 125
forces, 122, 122f, 125
    balanced, 123, 123f, 126, 127
    buoyancy, 134, 134f, 135
    and motion, 150–151
    reaction, 126, 127
    representing, 124–125, 124f, 125f
    resistance, 126, 127
    unbalanced, 127, 128–129
    as vectors, 122
freezing point, of water, 208, 208f
frequency, 6, 7
frequency table, 6, 6t
fresh water, 100, 101, 204–205, 204f, 212
friction, 130–131, 130f, 132, 133
fruits, 67, 67f
fuels, 218, 219
fungi, 27

G
galaxies, 175, 175f, 196, 196f, 197
galaxy clusters, 197, 197f
gallium, 99
gases
    in air, 216–217
    concentration, 156, 157
    diffusion in alveoli, 156, 156f
    percentages, in dry air, 214, 214t
gas exchange, 152, 156–157, 172–173
gas giant, 180, 181
'glow' foods, 67, 71
glucose, 20, 21, 24, 218
'go' foods, 66, 70
graphs, 9, 9f
    distance-time, 140–141, 140f, 141f
    velocity-time, 144, 144f
gravity, 182–183, 182f
green plants, 24, 26, 26f
Group 17 elements, 47, 47f
Group 1 metals, 46, 46f
groups, 43
    of Periodic Table, 46–47
Groups 3–12, of Periodic Table, 47
'grow' foods, 66, 66f, 70

H

hair hygrometer, 17, 17f
Halley's Comet, 194, 195
halogens, 47
heart muscle cells, 161, 161f
heat
conductor of, 44, 45
effect on substances, 96–97
heat radiation, 212, 213
herbivores, 19, 27
human nutrition, 64–65
hydrocarbons, 102, 103
hydrochloric acid, 74, 94
hydrogen, 46
hydrogen chloride, diffusion of, 53, 53f
hydrogen peroxide, 95, 95f
hydrometers, 150–151, 151f
hygrometer, hair, 17, 17f

I

iceberg, 135, 135f
impure substance, 91
independent variables, 12, 13
inert, 216, 217
ingestion, 65
inhaling, 158, 158f, 159
intercostal muscles, 158, 159
internal (cell, tissue) respiration, 160, 161
intestines, 74, 74f
large, 74f, 75
small, 74f, 75
iodine, 21
iodine solution, 78, 79
iron, and sulfur, 106–107, 106f
iron sulfide, 106–107, 106f, 107f
irregular galaxy, 196, 197
irreversible change, 97
isotopes, 50–51, 51f

J

joules, 166, 167
Jupiter, 180, 180f

K

kinetic energy, 41, 41f, 52
km/h (kilometres per hour), 137

L

lactic acid, 165
large intestine, 74f, 75
leaf
cells, 18
structure of, 20–21, 20f
light
materials react with, 186
sources of, 186, 188
limestone, 219
line of best fit, 10, 10f, 11
lithium, 46
luminous light sources, 186, 187
lunar eclipse, 192–193, 193f
lungs, 154, 154f, 155
air pressure in, 158
gas exchange in, 152, 156–157

M

magnesium, 104, 104f
magnetic, 106, 107
magnetising, 90f
magnitude, 124, 125
and direction, 142–143, 142f, 143f
Mars, 177, 177f, 178–179, 178f, 179f, 184
Mars Explorer project, 179, 179f
mass number, 48, 49, 50, 51
and isotopes, 50–51, 51f
menu card, examples of, 89, 89f
mercury, 33, 34
Mercury (planet), 178, 178f
metallic elements, 40f
metalloids, 44
metals, 45f
Group 1, 46, 46f
positions in Periodic Table, 44f
properties of, 44–45
meteorite, 194, 195
meteoroids, 194, 195
meteors, 178, 179, 194, 194f, 195
methane, 180, 181
methylmercury, 33
Milky Way, 196, 197
Minamata disease, 32–33
minerals, 67

mitochondria, 161, 161f

mixtures, 100–103

    air, 101

    alloy, 102, 102f, 103

    characteristics of, 103t

    crude oil, 102, 102f, 103

    filtration, 109, 109f

    iron and sulfur, 106–107, 106f

    as physical process, 100, 101

    physical separation, 108–109

    separation, 120, 120f

    water, 100, 100f

moon, 177, 177f, 184–185, 184f, 190. *see also* lunar eclipse

    landings, 185

    orbit of, 184, 184f

    phases of, 184–185, 184f

motion, 123, 123f

    forces and, 150–151

mouth, 72, 72f, 73

m/s^2, 144

m/s (metres per second), 136, 137

N

natural satellites, 176, 177, 184

natural springs, 213

Nature Conservancy of Jamaica, 38

Neptune, 181, 181f

neutrons, 48, 49

Newton, Isaac, 182, 183

newtons (N), 124, 125

nitrogen, 156, 216

non-luminous light sources, 186, 187

non-metallic elements, 40f, 44, 44f

non-metals, 45f

    positions in Periodic Table, 44f

    properties of, 45

nose, 154, 154f, 155

nucleus, 48

nutrition, 64–65

    in green plants, 18

    human, 86–87

    processes in, 65

nutritional information, 68, 68f, 69

O

omnivores, 19

opaque material, 186, 186f, 187

osmosis, 41, 54–55, 60

oxygen, 22, 24, 25, 152, 156, 157, 164, 215, 216, 216f

P

patterns, data, 10, 10f, 11

penumbra, 188, 189

peptides, 74

period, 43

Periodic Table, 42–43, 43f

    atomic number and, 50, 50f

    attractive version, creation of, 62–63

    groups of, 46–47

    Groups 3–12 of, 47

    position in, 44, 44f

petrol, 105

photosynthesis, 18–19, 18f, 22–23, 22f, 215

    and energy relationships, 38–39

    products, importance of, 24

    testing, 21

physical changes, 76, 77, 90, 90f, 120–121

    chemical changes *vs.*, 96–99

    magnetising, 90f

    melting, 97

physical process, 100, 101

physical properties, of matter, 92–93

    qualitative, 93

    quantitative, 93

pie chart, 6–7, 6f, 7f, 8, 8f

    and graphs, 9, 9f

planets, 174, 174f, 176, 176t, 177. *see also* specific planets

    inner four, 178–179, 178f

    outer four, 180–181, 180f–181f

point light source, 188, 188f, 189

polymer, 20, 21

potassium, 46

precipitation, 212, 213

predator, 31

prey, 30, 31

primary consumers, 26, 27

producers, 19, 26
products, 104, 105
proteases, 74
proteins, 66, 67
    testing for, 80, 80f
protons, 48, 49
Proxima Centauri, 198
pure substance, 91, 91f

Q
qualitative physical properties, 93
quantitative information, 8, 9
quantitative physical properties, 93

R
racing cars, 133, 133f
rain, 204f, 205, 212
Ramsay, William, 11
random motion, 52, 52f
reactants, 104, 105
reaction forces, 126, 127
red planet. see Mars
relative density, 150
replenished, 24, 25
reservoirs, 210, 211, 212
residue, 109
resistance forces, 126, 127
respiration, 215
    cell, 24, 25, 152, 160–161, 160f
    and gas exchange, 152–153, 172–173
    internal (cell, tissue), 160, 161
    process of, 215
respiratory system
    and breathing, 152
    and gas exchange, 152, 172–173
    parts of, 154
    structure of, 152, 154–155, 154f
reversible change, 97
rice, 66f
rocket, 143, 143f
root vegetables, 24
rust/rusting, 90, 90f, 214, 215

S
saliva, 73
satellites, natural, 176, 177, 184, 185
Saturn, 180, 180f

scalar quantities, 124, 124t, 125
scavengers, 27f
sea shells, 219, 219f
sea water, 100, 100f, 101, 204
secondary consumers, 26, 27
sectors, 8
sedimentation, 225, 225f
semi-conductor, 45
semi-metals, 44, 45
    positions in Periodic Table, 44f
    properties of, 45
settling tank, 210, 211
shadow, 186, 186f, 187
    full, 188, 188f
    partial, 188, 188f
shelf life, 216, 217
sieving, 109
silicon, 44, 45, 45f
simple sugars, 79
    testing for, 78, 78f
sink, 134–135, 134f
skeletal muscle cells, 161, 161f
slurry, 104, 105
small intestine, 74f, 75
snakes, 77, 77f
solar cells, 206, 207
solar eclipse, 190–191, 190f
Solar System, 174, 174f, 176–177
    comets, 194–195, 194f
    inner four planets, 178–179, 178f
    meteors, 194, 194f, 195
solubilities, 115
solvent, 115
space, distance in, 198–199
space science, 174–175, 202–203. see also
    Solar System
specific gravity, 150
speed, 136–137
    average, 138, 139
    calculation, 138–139, 138f
speedometers, 137, 137f
spiral galaxy, 196, 196f, 197
spirometers, 172–173, 172f, 173f
starch, 20, 21, 24, 79
    testing for, 78, 78f

stars, 175, 175f, 196

STEAM (Science, Technology, Engineering, Art and Mathematics) activities, 16–17

steel, 102, 102f, 103

stoma, 20, 21

stomach, 74, 74f, 75

streamlined, 132, 132f, 133

sub-atomic particles, 48–49, 48f, 49t

    calculating numbers of, 50–51

sulfur, iron and, 106–107, 106f

Sun, 143, 143f, 175, 176, 182, 182f, 196. *see also* solar eclipse

    diameter of, 190

    energy, and photosynthesis, 22

super cluster, 197

surface temperature, 178, 179

surface tension, 209

sweet potatoes, 60, 60f

T

teeth, 72, 72f, 73

time, relationship between distance and speed, 138–139

trachea, 154, 154f, 155

transition elements, 47, 47f

translucent material, 186, 186f, 187

transparent material, 186, 186f, 187

transpiration, 212, 213

trend, 10, 11

U

umbra, 188, 189

unbalanced forces, 127, 128–129

Universe, 198–199, 198f

upthrust, 134, 134f, 135

Uranus, 180, 180f

V

variables. *see also* specific types

    defined, 12, 13

    examples, 12

    in scientific experiments, 12–13

variegated leaves, 23, 23f

vectors

    forces as, 122

    quantities, 124, 124t, 125

vegetables, 67, 67f

velocity, 124

    displacement and, 142–143

    rocket, 143, 143f

velocity-time graphs, 144, 144f

Venus, 178, 178f

vitamins, 67

volatility, 110, 111

vultures, 27f

W

water, 64, 157, 160

    boiling point of, 208, 208f, 209

    conservation, 211

    density of, 150

    distilled, 91, 91f, 100, 101

    drinking, purification of, 210, 210f–211f

    and Earth's atmosphere, 204–205, 204f, 205f, 224–225

    freezing point of, 208, 208f

    fresh, 204–205, 204f, 212

    mixtures, 100, 100f

    molecules, 209

    properties of, 208–209

    purification plant, 224, 224f

    sea, 100, 100f, 101, 204

    sources of, 206

    testing for, 208–209

    uses of, 206

water cycle, 212–213, 212f

water resistance, 132, 133

weight, 126, 127

welding, 216, 217

wells, 206, 207

whales, and breathing, 153, 153f

windlass, 207, 207f

worms, 27, 27f

Acknowledgements

The publishers wish to thank the following for permission to reproduce photographs. Every effort has been made to trace copyright holders and to obtain their permission for the use of copyright materials. The publishers will gladly receive any information enabling them to rectify any error or omission at the first opportunity.

p6-7: Monkey Business Images/Shutterstock, p6: Monkey Business Images/Shutterstock, p12: MARTYN F. CHILLMAID/SCIENCE PHOTO LIBRARY, p13: Stefano Gilera/Cultura/Getty Images, p13: Mona Makela/Shutterstock, p13: LutsenkoLarissa/Shutterstock, p15: ANDREW LAMBERT PHOTOGRAPHY/SCIENCE PHOTO LIBRARY, p16: Kari Rene Hall / Contributor / Getty Images, p17: DAVID PARKER/SCIENCE PHOTO LIBRARY, p17: DAVID PARKER/SCIENCE PHOTO LIBRARY, p17: Eduard Gebel/Shutterstock, p17: Vitaly Korovin/Shutterstock, p17: kariphoto/Shutterstock, p18-19: Greg Brave/Shutterstock, p18: KKulikov/Shutterstock, p19: pitsanu suanlim/Shutterstock, p20: pukach/Shutterstock, p22: Afoto6267/Shutterstock, p23: Hiroya Minakuchi/Getty Images, p23: Kwanbenz/Shutterstock, p24: Dorling Kindersley/Getty Images, p24: ANDREW LAMBERT PHOTOGRAPHY/SCIENCE PHOTO LIBRARY, p26: Adam Hurley/EyeEm/Getty Images, p26: Elena Elisseeva/Shutterstock, p27: Flirt/P. Bauer/Corbis/Alamy, p27: Bernhard Richter/iStockphoto, p27: Javarman/Shutterstock, p28: FRANS LANTING, MINT IMAGES / SCIENCE PHOTO LIBRARY, p28: Studiotouch/Shutterstock, p37: Binh Thanh Bui/Shutterstock, p38: Anna Hoychuk/Shutterstock, p39: CH Collection/Alamy Stock Photo, p39: D. Kucharski K. Kucharska/Shutterstock, p39: Tao Jiang/Shutterstock, p39: PRILL/Shutterstock, p39: weter 777/Shutterstock, p40-41: Jurik Peter/Shutterstock, p40: Malashevska Olena/Shutterstock, p40: ppart/Shutterstock, p40: Tom Grundy/Shutterstock, p40: Bozena Fulawka/Shutterstock, p41: Nico Tondini/age fotostock/Alamy Stock Photo, p43: Humdan/Shutterstock, p44: concept w/Shutterstock, p45: Phil Degginger/Alamy Stock Photo, p45: an Miles-Flashpoint Pictures/Alamy Stock Photo, p45: bagi1998/Getty Images, p45: lucentius/E+/Getty Images, p45: SCIENCE PHOTO LIBRARY, p45: ppart/Shutterstock, p45: pedphoto36pm/Shutterstock, p45: Tyler Boyes/Shutterstock, p45: Bokeh Art Photo/Shutterstock, p46: MARTYN F. CHILLMAID/SCIENCE PHOTO LIBRARY, p47: Dorling Kindersley/UIG/SCIENCE PHOTO LIBRARY, p47: SCIENCE PHOTO LIBRARY, p48: Time Life Pictures/Mansell/The LIFE Picture Collection/Getty Images, p60: MARTYN F. CHILLMAID/SCIENCE PHOTO LIBRARY, p60: Hong Vo/Shutterstock, p61: vm2002/Shutterstock, p63: VIEW Pictures Ltd / Alamy Stock Photo, p64-65: Syda Productions/Shutterstock, p64: Burke/Triolo Productions/Getty Images, p64: Sean Nel/Shutterstock, p66: Elena Elisseeva/Shutterstock, p66: Bonchan/Shutterstock, p66: Matthew Bechelli/Shutterstock, p67: Fotofreaks/Shutterstock, p68: Elena Schweitzer/Shutterstock, p70: Jake Lyell/Alamy, p70: Dallas Stribley/Getty Images, p70: Blacqbook/iStockphoto/Getty Images, p70: Avava/Shutterstock, p71: Photofusion/Getty Images, p77: Sugar0607/iStockphoto, p77: Gayvoronskaya Yana/Shutterstock, p77: Eric Isselee/Shutterstock, p78: ANDREW LAMBERT PHOTOGRAPHY/SCIENCE PHOTO LIBRARY, p78: Martyn F. Chillmaid/Science Source, p79: SATURN STILLS/SCIENCE PHOTO LIBRARY, p86: BSIP/UIG/Getty Images, p88: Clynt Garnham Food & Drink / Alamy Stock Photo, p88: urbanbuzz/Shutterstock, p88: urbanbuzz/Shutterstock, p89: Answer Production/Shutterstock, p89: Decorwithme/Shutterstock, p89: mountain beetle/Shutterstock, p90-91: Jag_cz/Shutterstock, p90: Ray Pfortner/Getty Images, p90: Sergio Foto/Shutterstock, p90: Erasmus Wolff/Shutterstock, p90: Lighttraveler/Shutterstock, p91: GIPhotoStock Z/Alamy Stock Photo, p98: Michael Krinke/iStockphoto, p98: MARTYN F. CHILLMAID/SCIENCE PHOTO LIBRARY, p98: Pavel Vakhrushev/Shutterstock, p100: D. Callcut/Alamy Stock Photo, p100: Kanate/Shutterstock, p101: CHARLES D. WINTERS/SCIENCE PHOTO LIBRARY, p102: MediaWorldImages / Alamy Stock Photo, p102: Markus Mainka / Alamy Stock Photo, p102: PAUL RAPSON/SCIENCE PHOTO LIBRARY/Getty Images, p102: Fat Jackey/Shutterstock, p103: Jason Bye / Alamy Stock Photo, p104: Helene Rogers/ Art Directors & TRIP/

Acknowledgements

Alamy Stock Photo, p104: LH Images/Alamy Stock Photo, p104: ANDREW LAMBERT PHOTOGRAPHY/SCIENCE PHOTO LIBRARY, p104: golubovystock/Shutterstock, p104: focal point/Shutterstock, p104: urbanbuzz/Shutterstock, p106: TREVOR CLIFFORD PHOTOGRAPHY/SCIENCE PHOTO LIBRARY, p106: MARTYN F. CHILLMAID/SCIENCE PHOTO LIBRARY, p106: CHARLES D. WINTERS/SCIENCE PHOTO LIBRARY, p106: MARTYN F. CHILLMAID/SCIENCE PHOTO LIBRARY, p106: LAWRENCE MIGDALE/SCIENCE PHOTO LIBRARY, p107: sciencephotos/Alamy Stock Photo, p113: David Lefranc/Corbis, p115: ANDREW LAMBERT PHOTOGRAPHY/SCIENCE PHOTO LIBRARY, p120: SPENCER GRANT/SCIENCE PHOTO LIBRARY, p121: G Allen Penton/Shutterstock, p122-123: PhilipYb Studio/Shutterstock, p122: Stocktrek Images, Inc./Alamy Stock Photo, p122: Stanley Chou/Getty Images, p122: Anna Baburkina/Shutterstock, p123: Terry Foster/Alamy Stock Photo, p123: FatCamera/E+/Getty Images, p123: Michael Steele/Getty Images, p126: Henry Beeker/Alamy Stock Photo, p126: Don Despain/Alamy Stock Photo, p126: Lucky Business/Shutterstock, p126: Mino Surkala/Shutterstock, p128: Frank Miesnikowicz/Alamy Stock Photo, p128: NASA, p128: ID1974/Shutterstock, p132: Maximilian Weinzierl/Alamy, p132: Gallo Images/Getty Images, p132: Zorandim/Shutterstock, p133: Stuart Hickling/Alamy, p133: FPG/Getty Images, p133: SuperStock/Getty Images, p133: Reuters, p135: topseller/Shutterstock, p135: Alones/Shutterstock, p136: Pete Saloutos/Getty Images, p136: Jose Luis Pelaez Inc/Getty Images, p137: Winai Tepsuttinun/Shutterstock, p138: David J. Green/Alamy, p138: Al Tielemans /Sports Illustrated/Getty Images, p139: FABRICE COFFRINI/AFP/Getty Images, p143: Ro-Ma Stock Photography/Getty Images, p143: MARK GARLICK/SCIENCE PHOTO LIBRARY, p145: Leo Mason/Leo Mason/Corbis, p148: mRGB/Shutterstock, p151: Alexandre Dotta/SCIENCE PHOTO LIBRARY, p152-153: Life science/Shutterstock, p153: Ian Shaw/Alamy Stock Photo, p153: Anthony Asael/Art in All of Us/Corbis, p153: Randimal/Shutterstock, p155: National Geographic Creative / Alamy Stock Photo, p156: Ugo Montaldo/Shutterstock, p157: Fiona Bailey/Stockim/Alamy, p159: Yon Marsh/Alamy, p161: MICROSCAPE/Science Photo Library, p161: Jose Luis Calvo/Shutterstock, p162: Bog Dan/Anadolu Agency/Getty Images, p162: Hugo Felix/Shutterstock, p164: Reynold Mainse/Design Pics/Corbis, p164: Lucian Coman/Shutterstock, p166: Science Museum/SSPL/Getty Images, p166: Davydenko Yuliia/Shutterstock, p166: Tobik/Shutterstock, p166: maradon 333/Shutterstock, p166: givaga/Shutterstock, p166: Nikola Bilic/Shutterstock, p166: Andrei Kholmov/Shutterstock, p170: Ekaterina_Minaeva/Shutterstock, p172: Boom/Alamy Stock Photo, p172: DR P. MARAZZI/SCIENCE PHOTO LIBRARY, p172: glenda/Shutterstock, p172: Janthiwa Sutthiboriban/Shutterstock, p174-175: IM_photo/Shutterstock, p174: NASA, p174: NASA, p175: NASA, p177: NASA, p178: NASA, p178: NASA, p178: NASA, p178: Tristan3D/Shutterstock, p179: NASA, p180: NASA, p180: NASA, p180: NASA, p181: NASA, p184: NASA, p184: shooarts/Shutterstock, p185: NASA, p186: Idea Images/Getty Images, p186: Lester Balajadia/Shutterstock, p186: Karin Hildebrand Lau/Shutterstock, p186: Goldnetz/Shutterstock, p190: Lekcej/iStockphoto, p191: Stocktrek Images/Getty Images, p191: Teekid/iStockphoto, p193: Worldswildlifewonders/Shutterstock, p193: Thomas Nord/Shutterstock, p194: Robert Mikaelyan/NASA, p196: NASA / S.Dupuis / Alamy Stock Photo, p196: ESA/Hubble & NASA Acknowledgeme, p196: Knut Lundmark/NASA, p197: NASA, J. English (U. Manitoba), S. Hunsberger, S. Zonak, J. Charlton, S. Gallagher (PSU), and L. Frattare (STScI), p197: Yganko/Shutterstock, p198: Science Photo Library/Alamy Stock Photo, p202: ES Tech Hacks and Pranks Archive / Alamy Stock Photo, p204-205: Dmitry Polonskiy/Shutterstock, p204: Dennis Hallinan/Alamy Stock Photo, p204: NASA, p205: studio23/Shutterstock, p206: THEERASAK TAMMACHUEN/Shutterstock, p208: sciencephotos/Alamy Stock Photo, p208: itor/Shutterstock, p208: Petr Malyshev/Shutterstock, p212: Josias Gob Studio inc/Moment/Getty Images, p213: Art Directors & TRIP / Alamy Stock Photo, p215: Sunny Forest/Shutterstock, p216: geogphotos/Alamy Stock Photo, p216: Lisa F. Young/Shutterstock, p217: Anton Prado PHOTO/Shutterstock, p219: Evlakhov Valeriy/Shutterstock, p222: John Griffiths/Shutterstock, p224: Kekyalyaynen/Shutterstock, p224: Sarayuth_W/Shutterstock, p225: Zaharia_Bogdan/Getty Images, p225: Keith Homan/Shutterstock.